Application of Machine Learning in Slope Stability Assessment

Zhang Wengang · Liu Hanlong · Wang Lin ·
Zhu Xing · Zhang Yanmei

Application of Machine Learning in Slope Stability Assessment

Science Press
Beijing

Springer

Zhang Wengang
School of Civil Engineering
Chongqing University
Chongqing, China

Wang Lin
Beijing Normal University
Zhuhai, China

Zhang Yanmei
Chongqing University
Chongqing, China

Liu Hanlong
Chongqing University
Chongqing, China

Zhu Xing
Chengdu University of Technology
Chengdu, China

ISBN 978-981-99-2755-5 ISBN 978-981-99-2756-2 (eBook)
https://doi.org/10.1007/978-981-99-2756-2

Jointly published with Science Press
The print edition is not for sale in China mainland. Customers from China mainland please order the print book from: Science Press.

This Springer imprint is published by the registered company Springer Nature Singapore Pte Ltd.
The registered company address is: 152 Beach Road, #21-01/04 Gateway East, Singapore 189721, Singapore

Foreword

The field of slope engineering encompasses slope stability analysis and design, movement monitoring, and slope safety management and maintenance. Based on the material composition, it can be sub-divided into soil slope, rock slope and soil–rock slope. Natural and man-made slope failures are complex phenomena and cause serious hazards in many countries of the world, particularly China. As a result, public and private property is damaged worth millions of dollars. It is necessary to understand the processes causing failure of slopes and prediction of its vulnerability for proper mitigation of slope failure hazards. Various attempts have been made to predict the stability of slope using both conventional methods such as limit equilibrium method, finite element method, finite difference method, and statistical methods. Artificial intelligence (AI) methods like artificial neural networks, genetic programming and genetic algorithms, and support vector machines, etc. are found to have better efficiency compared to statistical methods. Based on the recent references, especially the publications by authors of current book, application of these machine learning (ML) techniques is becoming more wide and common.

This book covers the scopes and methods mentioned above, and it tries to cope with slope stability analysis and deformation calculation practice, implemented via the recent publications from the authors in reputable journals. It comprises 11 chapters. Chapter 1 overviews the slope stability analysis methods, including the theoretical solutions, numerical simulations, physical experimentations as well as the in-situ monitoring methods and the machine learning approaches. Chapter 2 introduces the applications of the main machine learning algorithms. Chapter 3 focuses on introduction of the smart in-situ monitoring and slope stability assessment based on two well-documented case histories. Chapter 4 presents the prediction of slope stability using ensemble learning techniques while Chap. 5 concentrates on landslide susceptibility research combining qualitative analysis and quantitative evaluation. Chapter 6 applies transfer learning (TL) to improve landslide susceptibility modeling performance while Chap. 7 performs displacement prediction of Jiuxianping landslide using gated recurrent unit (GRU) networks. Chapter 8 carries out efficient seismic stability analysis of slopes subjected to water level changes using gradient boosting algorithms. Chapter 9 presents efficient reliability analysis of slopes in spatially variable soils using XGBoost. Chapter 10 conducts efficient time-variant reliability analysis of Bazimen landslide in the Three Gorges Reservoir Area using XGBoost and LightGBM algorithms while Chap. 11 provides the summary of the book and put up with the future work recommendation.

I am pleased to witness the support of National Fund for Academic Publication in Science and Technology, MOST and publication of this monograph and hope that the authors would bring us more findings in application of machine learning in slope engineering when they reprint it.

August 2022

Xiangsheng Chen
Academician of Chinese Academy
of Engineering
Guangzhou, China

Preface

The so-called Fourth Paradigm has been boomingly developed during the past two decades, in which large quantities of observational data are available to scientists and engineers. Big data is characterized by the rule of the five Vs: Volume, Variety, Value, Velocity, and Veracity. The concept of big data naturally matches well with the features of geoengineering and geoscience. Large-scale, comprehensive, multi-directional, and multi-field geotechnical data analysis is becoming a trend. On the other hand, soft computing (SC), machine learning (ML), deep learning (DL), and optimization algorithm (OA) provide the ability to learn from data and deliver in-depth insight into geotechnical problems. Researchers use different SC, ML, DL, and OA models to solve various problems associated with geoengineering, including the slope engineering. Consequently, there is a need to extend the slope stability research with big data research through integrating the use of SC, ML, DL, and OA techniques.

This book focuses on the state-of-the-art and application of ML algorithms in slope stability assessment and deformation prediction. Various ML approaches are firstly concisely introduced. Then their representative applications in slope engineering are presented, including the ensemble learning (EL), transfer learning (TL), gated recurrent unit (GRU), and XGBoost and LightGBM algorithms. The authors also provided their own thoughts learnt from these applications as well as work ongoing and future recommendations. This book aims to make a comprehensive summary and provide fundamental guidelines for researchers and engineers in the discipline of geotechnical slope engineering or similar research areas on how to integrate and apply ML methods.

In the process of writing this book, the authors received enthusiastic help and invaluable assistance from many people, which is deeply appreciated. The first author would like to express his special thanks to Mr. Tang Libin, Mr. Wu Chongzhi, Mr. Liu Songlin, Mr. Gu Xin, Miss Li Hongrui, and especially Miss Zhou Xiaohui, for their help in editing the book draft and professional page proofreading. The first author is grateful to his Ph.D. supervisor, Prof. Goh Anthony Teck Chee of Nanyang Technological University, for years of supervision and help, as well as the communication via emails, even though he left Singapore for almost seven years. Professor Goh leads him toward early career development.

The support from the National Fund for Academic Publication in Science and Technology (MOST), National Key R&D Program of China (Grant No. 2019YFC1509605); the National Natural Science Foundation of China (Grant Nos. 52078086 and 52008058); MOST High-end Foreign Expert Introduction program (Grant Nos. DL2021165001L, G20200022005, and G2022165004L) Program of Distinguished Young Scholars, Natural Science Foundation of Chongqing, China (Grant No. cstc2020jcyj-jq0087); and Cooperation projects between universities in Chongqing and affiliated institutes of the Chinese Academy of Sciences (Grant No. HZ2021001) is greatly acknowledged.

Finally, the first author would like to thank his family for their love, understanding, and support.

He would also like to end the preface via the definition of slope stability and an old Chinese saying:

Slope stability refers to the condition that an inclined slope can withstand its own weight and external forces without experiencing displacement.

Let it ripple and I stand still (任他波涛汹涌, 我自岿然不动).

Zhang Wengang

Chongqing, China
August 2022

Contents

About the Authors

Dr. Zhang Wengang is currently full professor in School of Civil Engineering, Chongqing University, China. His research interests focus on underground engineering, slope engineering, bio-inspired geotechnics, as well as big data and machine learning in geotechnics and geoengineering. He is now the members of the ISSMGE TC304 (Reliability), TC309 (Machine Learning), TC219 (System Performance of Geotechnical Structures) and TC222 (Digital Twin). He also serves *Geoscience Frontiers* as Associate Editor, Editorial member for *Georisk*, *Journal of Rock Mechanics and Geotechnical Engineering* as well as Underground Space. Dr Zhang has been selected as 2021 Highly cited Chinese Scholars and the World's Top 2% Scientists for years 2019 and 2020. He won the 2019 Computers and Geotechnics Sloan Outstanding Paper Award and 2021 Underground Space Outstanding Paper Award.

Prof. Liu Hanlong is the deputy vice President of Chongqing University and Chair professor in Geotechnical Engineering of the school. He obtained his Bachelor degree from Zhejiang University in 1986 and Ph.D. degree from Hohai University in 1994, China. He was awarded the Chang Jiang Scholar Award by the Ministry of Education in 2007 and the Outstanding Young Scholar Award by the National Science Foundation of China in 2008. He is a member of ISO committee for the international standard on Seismic Action for Designing Geotechnical Works (ISO23469) and editor-in-chief of national technical code for ground treatment of hydraulic fill (GB/T51064-2015).

Dr. Wang Lin is currently an associate research fellow in School of National Safety and Emergency Management, Beijing Normal University, China. His research interests focus on geotechnical reliability analysis, landslide risk assessment, as well as machine learning and its applications in geotechnical and geological engineering. He served as an Early Career Editorial Board Member for the *Georisk*. He has published many academic papers and received several research projects.

Dr. Zhu Xing is currently associated professor in College of Computer Science and Cyber Security (Oxford Brookes College), Chengdu University of Technology, China. His research interests focus on intelligent Wireless Sensor Networks (iWSNs) for monitoring and early warning of geohzards, big data, and machine learning in engineering geology. He is also the research member of State Key Laboratory of Geohazard Prevention and Geoenvironment Protection, China. He developed a series of smart sensors and Internet of Things techniques for landslide/rockfall monitoring and solved the early warning difficulties and problems of sudden landslides fundamentally. As a core member, he has won the second prize of the National Science and Technology Progress Award in 2019.

Dr. Zhang Yanmei is currently Associate Professor in College of Aerospace Engineering, Chongqing University, China. Her research interests focus on fatigue and failure assessment of engineering structures, fracture analyses of polymeric composites, and applications of machine learning technique in geoengineering. She is the member of Chinese Society of Theoretical and Applied Mechanics.

Symbols and Abbreviations

The following list of symbols and abbreviations is provided for ease of reference for those symbols and abbreviations that are most frequently used in this book. It is therefore not exhaustive. Nonetheless, all symbols and abbreviations used in this book are defined at their first mention in the main text.

Symbols

α	Tangential angle
v_i	Daily average velocity
$\bar{v}$	Overall average velocity
t_i	Current monitoring time period
Δv	Velocity increment
$\widehat{y}_i^{(t)}$	Final tree model
$\widehat{y}_i^{(t-1)}$	Previous tree model
$f_t(x_i)$	Newly generated tree model
$Obj^{(t)}$	Objective function
$l(y_i, \widehat{y}_i^{(t)})$	Loss function
$\Omega(f_i)$	Regularization term
T	The number of leaves
w	The weights of the leaf
D_S	Source domain
T_S	Learning task
$\mathbf{r}_t$	Reset gate
$\mathbf{z}_t$	Update gate
$\widetilde{\mathbf{h}}_t$	Hidden layer
$x_{\min}$	The minimum value
$x_{\max}$	The maximum value
R^2	Coefficient of determination

b	Bias factor
P	Prior value
Y_{σ_j}	Label value
K_h	Horizontal seismic coefficient
P_f	Probability of failure
H	Total head
S_e	Effective degree of saturation
θ_r	Residual volumetric water content
θ_s	Saturated volumetric water content
ψ	Matric suction
τ	Unsaturated shear strength
σ	Total normal stress
$\rho_{i,j}$	Correlation coefficient

Abbreviations

ACF	Autocorrelation function
AI	Artificial intelligence
ANN	Artificial neural network
AUC	The area under the ROC curve
BF	Basic function
CatBoost	Categorical boosting
CBR	Case-based reasoning
CNN	Convolutional neural network
COV	Coefficient of variation
DEM	Digital elevation model
DL	Deep learning
DT	Decision tree
EFB	Exclusive feature bundling
FN	False negative prediction
FP	False positive prediction
FS	Factor of safety
GBDT	Gradient boosting decision tree
GBM	Gradient boosting machine
GIS	Geographic information system
GNSS	Global navigation satellite system
GOSS	Gradient-based one-side sampling
GRU	Gated recurrent unit
GS	Grid search
HAILS	Human activity intensity of land surface
IM	Index of moisture
LEWS	Landslide early warning system
LightGBM	Light gradient boosting machine

LR	Logistic regression
LSM	Landslide susceptibility mapping
LSTM	Long short-term memory
MAPE	Mean absolute percentage error
MARS	Multivariate adaptive regression splines
MCS	Monte carlo simulation
ML	Machine learning
NDVI	Normalized difference vegetation index
PDF	Probability density function
PSO	Particle swarm optimization
RF	Random forest
RFR	Random forest regression
RL	Reinforcement learning
RMSE	Rooted mean squared error
RNN	Recurrent neural network
ROC	Receiver operation characteristic
RS	Random search
RSM	Response surface method
RT	Regression tree
SECF	Single exponential autocorrelation function
SHAP	Shapley additive explanation
SKLGP	State Key Laboratory and Geohazard Prevention and Geoenvironment Protection
SOF	Scale of fluctuation
SOM	Self-organizing map
SS	Subset simulation
SVM	Support vector machine
SWCC	Soil–water characteristic curve
TGRA	Three gorges reservoir area
TL	Transfer learning
TN	True negative prediction
TP	True positive prediction
TWI	Topographic wetness index
WCA	Warmup-cosine annealing
XGBoost	Extreme gradient boosting

Chapter 1
Overview

1.1 Slope Stability Analysis Methods

1.1.1 Theoretical Solutions

Generally, the theoretical solutions of slope stability can be divided into qualitative and quantitative evaluations.

The qualitative evaluation methods mainly include the engineering geological analogy method and the stereographic projection method. Specifically, the various influential factors can be considered comprehensively by the engineering geological analogy method. And then, the stability and development trend of slopes can be predicted quickly (Kilburn and Petley 2003; Yu et al. 2014; Sun et al. 2014; Gao 2015; Wang et al. 2020f). In addition, the stereographic projection method is widely used in the preliminary judgment of unstable slopes (Tomas et al. 2012; Jia et al. 2013; Zhou et al. 2017). However, due to the difference in slopes' geological conditions, the accuracy of qualitative evaluation methods depends on the experience of the researchers.

The limit equilibrium analysis is the primary method to analyze the slope stability quantitatively. After considering the translational or rotational movement with an assumed or specific potential slip surface, this method is used to investigate the slope stability that tends to be failure under the influence of gravity (Zhu et al. 2003; Asadollahi and Tonon 2012). The limit equilibrium method can be feasible and straightforward to obtain precise results. After continuous revision and improvement, this method is the most commonly used to determine the mechanical state of slopes (Deng et al. 2016; Nilsen 2017; Chen et al. 2020b; Tonon 2020).

Since many factors are affecting the slope stability, mathematical methods are adapted to determine the stability quantitatively or semi-quantitatively, such as dissipation theory (Qin et al. 2019), chaos theory (Gao 2013), stochastic theory (Mitchell and Hungr 2017), fuzzy theory (Daftaribesheli et al. 2011), gray system theory (Liu and Cao 2015), catastrophe theory (Xia et al. 2015; Wang et al. 2020a), and so on.

Z. Wengang et al., *Application of Machine Learning in Slope Stability Assessment*,
https://doi.org/10.1007/978-981-99-2756-2_1

These theoretical methods can help us better understand the evolution process to a certain extent and provide a reference for the slope stability.

1.1.2 Numerical Simulations

The numerical simulations can obtain the slopes' stress or strain distribution and reproduce the dynamic failure. With the development of computers, software and computational mechanics theory, a variety of numerical methods have been applied to slope stability analysis. When performing numerical calculations, the finite element method (Clough and Woodward 1967; Griffiths and Lane 1999; Bendezu et al. 2017; Gao et al. 2019; Wang et al. 2020b; Gu et al. 2020; Chen et al. 2020a; Zhang et al. 2021), discrete element method (Cundall and Strack 1979; Wang et al. 2020c, e), and the discontinuous deformation analysis method (Shi 1988) are commonly used. These numerical simulation methods have different operating mechanisms and applicable scopes. For example, the finite element method is widely applied to analyze continuous deformation (Jiang and Magnan 1997; Lim et al. 2017; Li et al. 2019).

For instance, Shen et al. (2012) used the powerful spatial analysis functions of Geographic Information System (GIS) such as automatic profile generation and gridding of the digital elevation model (DEM) in the preliminary treatment of numerical modeling. And they further developed a program for transforming GIS data to a commercial finite difference code FLAC. Xing et al. (2017) simulated the post-failure behavior of the rock avalanche triggered by the Wenchuan earthquake in the Wenjia valley. Dadashzadeh et al. (2017) proposed an integrated methodology for probabilistic numerical modeling of rock slope stability. Based on the response surface method, this method is used to develop an explicit performance function from numerical simulations. The implementation of the proposed methodology is performed by considering a large potential rock wedge in Sumela Monastery, Turkey. Through trial and error, Li et al. (2017) found that a combination of the frictional model and Voellmy model can provide the best performance in simulating the rock avalanche triggered by the Lushan earthquake. Scaringi et al. (2018) adapted various models (PFC, MatDEM, MassMov2D, Massflow) to reproduce the Xinmo landslide and simulate the kinematics and runout of the potentially unstable mass. Liu et al. (2019) studied the dynamic collapse based on the block discrete element method. Bi et al. (2019) used discrete element methods to study the effects of the configuration of a baffle-avalanche wall system on rock avalanches in Tibet Zhangmu.

Notably, it is important to determine the geological model, mechanical model, and mechanical parameters during the numerical simulation. And these settings are the key factors to ensure the simulation results' reliability and validity (Tang 1997; Stead et al. 2006; Stead and Wolter 2015; Sarfaraz and Amini 2020).

1.1.3 Physical Experimentations

Based on the principle of similarity and field survey results, physical model experiments are used to establish a simulation model (Adhikapy and Dyskin 2007; Chen et al. 2019; Gu et al. 2021). Under certain conditions, the physical model experiments can predict or reproduce the evolution of dangerous rock masses.

As for the soil slope, Wu et al. (2015) studied the rainwater infiltration into the slope and the slope failures induced by artificial rainfall with different initial conditions. They divided the soil slope failures into three types: overall sliding failure, partial sliding failure, and flow slide. Abdoun et al. (2018) performed a small-scale centrifuge test of a sloping deposit to study the reaction and liquefaction potential of sloped deposits experimentally and established the repeatability of experimental data for numerical simulations. Morse et al. (2017) measured increasing strain localization in unsaturated slope cuts prior to abrupt failure to experimentally validate a recently developed theory for predicting the stability of cut slopes under unsaturated conditions. To investigate the failure mode and triggering mechanisms of shallow loess landslides induced by rainfall, Sun et al. (2019) conducted three categories of indoor physical model experiments of a loess slope with and without a vertical joint under different rainfall patterns. To quantitatively analyze the progressive failure of the slope, Darban et al. (2019) carried out a couple of small-scale experiments on slopes reconstituted with unsaturated pyroclastic soils that were subjected to continuous rainfall.

As for the rocky slope, Huang et al. (2007) studied the effects of slope features on the rock stopping position, the moving time, and the moving track of different rock block shapes. Bowman et al. (2012) investigated whether dynamic fragmentation can lead to more significant runout or spreading of physical model rock avalanches. Feng et al. (2012) investigated the failure mechanism of the Jiweishan landslide through the geotechnical centrifugal model experiment. Their results reproduced the driving-blocks-and-key-blocks mode of apparent dip. To analyze the Jianchuandong dangerous rock mass in the Three Gorges Reservoir area, Wang et al. (2020c) devised a new hydraulic coupling mechanism that represents the mechanical environment of the base rock mass in the deterioration zone. Their model indicates that the damage can magnify the effective stress experienced by the base rock mass and lead to the nonlinear acceleration of the deterioration of the dangerous rock mass. Chen and Orense (2020) conducted laboratory experiments to release dry rigid blocks on an inclined chute to investigate the mechanisms involved in the downslope motions of granular particles.

However, the cost of physical model experiments is generally high, and the experimental conditions are difficult to control.

1.2 Remote Monitoring Methods

Many methods are available for monitoring unstable slopes. The remote location of many unstable slope or landslide in complex mountain areas requested a need for monitoring system that can be accessed remotely and provide real-time monitoring results for risk management. Advances in electronic instrumentation and wireless sensor network can provide an effective way for monitoring unstable or potentially unstable slopes remotely. Slope stability monitoring involves selecting certain parameters and observing how they change with time. Figure 1.1 shows the framework of wireless sensor network for monitoring unstable slope. Many parameters in slope evolution can be measured automatically by field sensors or instrumentations, and the data is collected by datalogger and/or wireless nodes and transmitted to the remote server through cellular phone network and/or long-range radio network. Remote users can access the near real-time monitoring time series data in anywhere through internet.

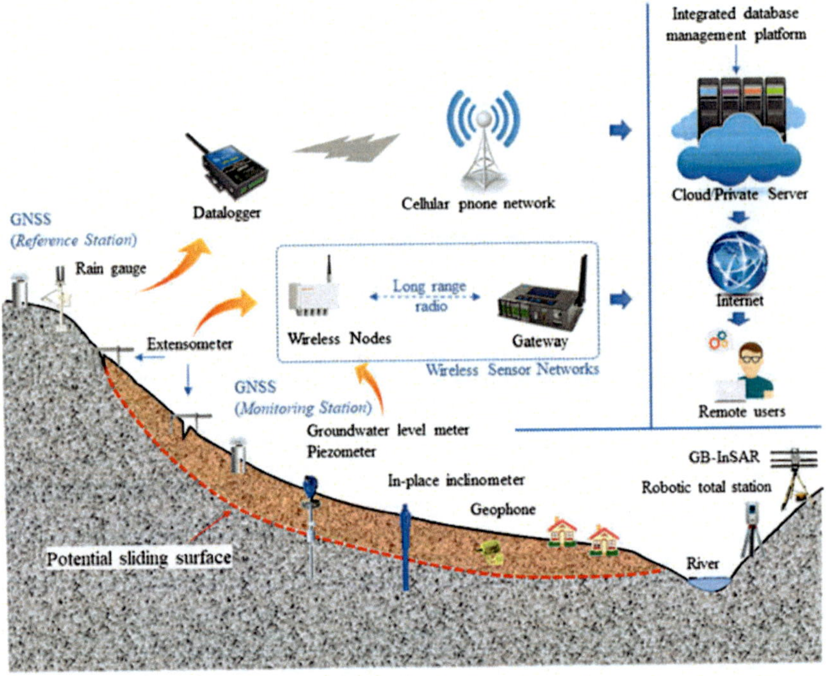

Fig. 1.1 Framework of wireless sensor network for monitoring unstable/potential slope

The critical data that are required from a slope monitoring program are deformation of slope and trigger factors.

(1) Displacement/movement monitoring

In-place inclinometers, extensometers, Global Navigation Satellite System (GNSS) can be used alone or in combination to monitor the displacement of slope movement (Simeoni and Mongiovì 2007, 2017a; Zhu et al. 2012; Xu et al. 2020). In-place inclinometers that is buried in the borehole can detect depth of potential sliding surface and deformations along different depth, while extensometer and GNSS, which are mounted at the ground surface, can monitor the ground surface movement of unstable slope. GNSS can achieve monitoring time series with millimeter accuracy of long-term slope deformation in three-dimensional space, while advanced low-cost extensometers can capture the rapid accelerating deformation of slope timely (Xu et al. 2020). Additionally, as a non-contact deformation monitoring method, robotic total station and GB-InSAR are more suitable for short-term monitoring of potential landslide for emergency management.

(2) Trigger factors monitoring

Rainfall and groundwater are the critical triggering factors contributing to slope movement (Zhu et al. 2017b; Miao et al. 2018; Zhang et al. 2020). Rain gauge sitting on the stable ground surface is used to monitoring rainfall parameters. Groundwater level meter or piezometer installed in the borehole case can detect the variation of groundwater level, which might be induced by rainfall and/or fluctuation of reservoir/river water level. Monitoring with automatic instrumentations can provide better data on how much rainfall is causing landslide and when they will occur. It also helps scientists learn how rainfall impacts the variation of groundwater level and movement of slope. Knowing the inner mechanisms of water on slope can lead to better tools for predicting the deformation of slope in the future (Zhu et al. 2018).

1.3 Machine Learning Approaches

1.3.1 What is Machine Learning

Machine learning (ML) is a branch of artificial intelligence (AI) and computer science which focuses on the use of data and algorithms to imitate the way that humans learn, and gradually improve its accuracy. It is an important component of the growing field of data science. Through the use of statistical methods, algorithms are trained to make classifications or predictions, uncovering key insights within data mining projects. These insights subsequently drive decision making within applications and businesses, ideally impacting key growth metrics. As big data continues to expand and grow, the market demand for data scientists will increase, requiring them to assist in the identification of the most relevant business questions and subsequently the data to answer them.

1.3.2 How Machine Learning Works

Generally, the learning system of a machine learning algorithm can be broken out into three main parts:

1. A Decision Process: In general, machine learning algorithms are used to make a prediction or classification. Based on some input data, which can be labeled or unlabeled, your algorithm will produce an estimate about a pattern in the data.
2. An Error Function: An error function serves to evaluate the prediction of the model. If there are known examples, an error function can make a comparison to assess the accuracy of the model.
3. A Model Optimization Process: If the model can fit better to the data points in the training set, then weights are adjusted to reduce the discrepancy between the known example and the model estimate. The algorithm will repeat this evaluation and optimization process, updating weights autonomously until a threshold of accuracy can be reached.

1.3.3 Machine Learning Methods

Machine learning classifiers fall into three primary categories: supervised machine learning, unsupervised machine learning, and semi-supervised learning. Since Chap. 3 would specifically introduce the applications of the main ML algorithm, and this section stops here for the three types.

1.3.4 What is Deep Learning

Deep learning (DL) attempts to mimic the human brain—albeit far from matching its ability—enabling systems to cluster data and make predictions with incredible accuracy.

It is a subset of machine learning, which is essentially a neural network with three or more layers. These neural networks attempt to simulate the behavior of the human brain—albeit far from matching its ability—allowing it to "learn" from large amounts of data. While a neural network with a single layer can still make approximate predictions, additional hidden layers can help to optimize and refine for accuracy.

Deep learning drives many AI applications and services that improve automation, performing analytical and physical tasks without human intervention. Deep learning technology lies behind everyday products and services (such as digital assistants, voice-enabled TV remotes, and credit card fraud detection) as well as emerging technologies (such as self-driving cars).

1.3.5 Deep Learning Versus Machine Learning

If deep learning is a subset of machine learning, how do they differ? Deep learning distinguishes itself from classical machine learning by the type of data that it works with and the methods in which it learns.

Machine learning algorithms leverage structured, labeled data to make predictions—meaning that specific features are defined from the input data for the model and organized into tables. This doesn't necessarily mean that it doesn't use unstructured data; it just means that if it does, it generally goes through some pre-processing to organize it into a structured format.

Deep learning eliminates some of data pre-processing that is typically involved with machine learning. These algorithms can ingest and process unstructured data, like text and images, and it automates feature extraction, removing some of the dependency on human experts. For example, let's say that we had a set of photos of different pets, and we wanted to categorize by "cat", "dog", "hamster", etc. Deep learning algorithms can determine which features (e.g., ears) are most important to distinguish each animal from another. In machine learning, this hierarchy of features is established manually by a human expert.

Then, through the processes of gradient descent and backpropagation, the deep learning algorithm adjusts and fits itself for accuracy, allowing it to make predictions about a new photo of an animal with increased precision.

Machine learning and deep learning models are capable of different types of learning as well, which are usually categorized as supervised learning, unsupervised learning, and reinforcement learning. Supervised learning utilizes labeled datasets to categorize or make predictions; this requires some kind of human intervention to label input data correctly. In contrast, unsupervised learning doesn't require labeled datasets, and instead, it detects patterns in the data, clustering them by any distinguishing characteristics. Reinforcement learning is a process in which a model learns to become more accurate for performing an action in an environment based on feedback in order to maximize the reward.

1.3.6 How Deep Learning Works

Deep learning neural networks, or artificial neural networks, attempt to mimic the human brain through a combination of data inputs, weights, and bias. These elements work together to accurately recognize, classify, and describe objects within the data.

Deep neural networks consist of multiple layers of interconnected nodes, each building upon the previous layer to refine and optimize the prediction or categorization. This progression of computations through the network is called forward propagation. The input and output layers of a deep neural network are called visible layers. The input layer is where the deep learning model ingests the data for processing, and the output layer is where the final prediction or classification is made.

Another process called backpropagation uses algorithms, like gradient descent, to calculate errors in predictions and then adjusts the weights and biases of the function by moving backward through the layers in an effort to train the model. Together, forward propagation and backpropagation allow a neural network to make predictions and correct for any errors accordingly. Over time, the algorithm becomes gradually more accurate.

The above describes the simplest type of deep neural network in the simplest terms. However, deep learning algorithms are incredibly complex, and there are different types of neural networks to address specific problems or datasets. For example,

convolutional neural networks (CNNs), primarily used in computer vision and image classification applications, can detect features and patterns within an image, enabling tasks, like object detection or recognition. In 2015, a CNN bested a human in an object recognition challenge for the first time.

Recurrent neural networks (RNNs) are typically used in natural language and speech recognition applications as it leverages sequential or times series data.

1.4 Organization of This Book

This book is arranged from regional slope stability assessment in Yunyang County to a site-specific reservoir slope, from static reliability analysis to long-duration dynamic reliability analysis, and from the prediction of conventional factor of safety to reliability index (or equivalently, failure probability). Accordingly, this book is divided into 11 chapters, and the main contents are summarized as follows:

This chapter overviews the slope stability analysis methods, including the theoretical solutions, numerical simulations, physical experimentations as well as the in-situ monitoring methods and the machine learning approaches.

Chapter 2 introduces the applications of the main machine learning algorithms, including the supervised learning, unsupervised learning, semi-supervised learning, reinforcement learning, deep learning, ensemble learning, etc., in slope engineering and landslide prevention.

Chapter 3 focuses on introduction of the smart in-situ monitoring and slope stability assessment based on two well-documented case histories.

Chapter 4 presents the prediction of slope stability using ensemble learning techniques.

Chapter 5 concentrates on landslide susceptibility research combining qualitative analysis and quantitative evaluation.

Chapter 6 applies transfer learning to improve landslide susceptibility modeling performance.

Chapter 7 performs displacement prediction of Jiuxianping landslide using gated recurrent unit (GRU) networks.

Chapter 8 carries out efficient seismic stability analysis of slopes subjected to water level changes using gradient boosting algorithms.

Chapter 9 presents efficient reliability analysis of slopes in spatially variable soils using XGBoost.

Chapter 10 conducts efficient time-variant reliability analysis of Bazimen landslide in the Three Gorges Reservoir Area using XGBoost and LightGBM algorithms.

Chapter 11 provides the summary of the book and puts forward the future work recommendation.

References

Abdoun T, Kokkali P, Zeghal M (2018) Physical modeling of soil liquefaction: repeatability of centrifuge experimentation at RPI. Geotech Test J 41:1

Adhikary DP, Dyskin AV (2007) Modelling of progressive and instantaneous failure of foliated rock slopes. Rock Mech Rock Eng 40(4):349–362

Asadollahi P, Tonon F (2012) Limiting equilibrium versus BS3D analysis of a tetrahedral rock block. Int J Numer Anal Meth Geomech 36(1):50–61

Bendezu M, Romanel C, Roehl D (2017) Finite element analysis of blast-induced fracture propagation in hard rocks. Comput Struct 182:1–13

Bi YZ, He SM, Du YJ, Sun XP, Li XP (2019) Effects of the configuration of a baffle-avalanche wall system on rock avalanches in Tibet Zhangmu: discrete element analysis. Bull Eng Geol Env 78(4):2267–2282

Bowman ET, Take WA, Rait KL, Hann C (2012) Physical models of rock avalanche spreading behaviour with dynamic fragmentation. Can Geotech J 49(4):460–476

Chen XY, Orense RP (2020) Investigating the mechanisms of downslope motions of granular particles in small-scale experiments using magnetic tracking system. Eng Geol 265:105448

Chen Z, Yang P, Liu H, Zhang W, Wu C (2019) Characteristics analysis of granular landslide using shaking table model test. Soil Dyn Earthq Eng 126:105761. https://doi.org/10.1016/j.soildyn.2019.105761

Chen L, Zhang W, Zheng Y, Gu D, Wang L (2020b) Stability analysis and design charts for over-dip rock slope against bi-planar sliding. Eng Geol 275:105732. https://doi.org/10.1016/j.enggeo.2020.105732

Chen F, Zhang R, Wang Y, Liu H, Thomas B, Zhang W (2020a) Probabilistic stability analyses of slope reinforced with piles in spatially variable soils. Int J Approximate Reasoning 122:66–79

Clough RW, Woodward RJ (1967) Analysis of embankment stresses and deformations. J Soil Mech Found Div 93:529–549

Cundall PA, Strack ODL (1979) A discrete numerical model for granular assemblies. Geotechnique 29(1):47–65

Dadashzadeh N, Duzgun HSB, Yesiloglu-Gultekin N (2017) Reliability-based stability analysis of rock slopes using numerical analysis and response surface method. Rock Mech Rock Eng 50(8):2119–2133

Daftaribesheli A, Ataei M, Sereshki F (2011) Assessment of rock slope stability using the Fuzzy Slope Mass Rating (FSMR) system. Appl Soft Comput 11(8):4465–4473

Darban R, Damiano E, Minardo A, Olivares L, Picarelli L, Zeni L (2019) An experimental investigation on the progressive failure of unsaturated granular slopes. Geosciences 9(2):63

Deng DP, Li L, Wang JF, Zhao LH (2016) Limit equilibrium method for rock slope stability analysis by using the Generalized Hoek-Brown criterion. Int J Rock Mech Min Sci 89:176–184

Feng Z, Yin YP, Li B, Yan J (2012) Centrifuge modeling of apparent dip slide from oblique thick bedding rock landslide. Chin J Rock Mech Eng 31(5):890–897 (in Chinese)

Gao XJ (2013) Bifurcation Behaviors of the two-state variable friction law of a rock mass system. Int J Bifurcation Chaos 23(11):1350184

Gao X, Liu H, Zhang W, Wang W, Wang Z (2019) Influences of reservoir water level drawdown on slope stability and reliability analysis. Georisk 13(2):145–453

Gao W (2015) Stability analysis of rock slope based on an abstraction ant colony clustering algorithm. Environ Earth Sci 73(12):7969–7982

Griffiths DV, Lane PA (1999) Slope stability analysis by finite elements. Geotechnique 49(3):387–403

Gu X, Wang L, Chen F, Li H, Zhang W (2020) Reliability analysis of slope stability considering temporal variations of rock mass properties. Comput Mater Continua 63 (1):263–281

Gu D, Liu H, Gao X, Huang D, Zhang W (2021) Influence of cyclic wetting-drying on the shear strength of limestone with a soft interlayer. Rock Mech Rock Eng 54:4369–4378

Huang RQ, Liu WH, Zhou JP, Pei XJ (2007) Rolling tests on movement characteristics of rock blocks. Chin J Geotech Eng 09:1296–1302 (in Chinese)

Jia JQ, Liu XH, Zhang X, Li XM (2013) Stability analysis of the wedge of rock slope and its programmed computation. J Comput Theor Nanosci 10(12):2902–2905

Jiang GL, Magnan JP (1997) Stability analysis of embankments: comparison of limit analysis with methods of slices. Geotechnique 47(4):857–872

Kilburn CRJ, Petley DN (2003) Forecasting giant, catastrophic slope collapse: lessons from Vajont, Northern Italy. Geomorphology 54(1–2):21–32

Li B, Xing AG, Xu C (2017) Simulation of a long-runout rock avalanche triggered by the Lushan earthquake in the Tangjia Valley, Tianquan, Sichuan, China. Eng Geol 218:107–116

Li Z, Hu Z, Zhang XY, Du SG, Guo YK, Wang JX (2019) Reliability analysis of a rock slope based on plastic limit analysis theory with multiple failure modes. Comput Geotech 110:132–147

Lim K, Li AJ, Schmid A, Lyamin AV (2017) Slope-stability assessments using finite-element limit-analysis methods. Int J Geomech 17(2):06016017

Liu DX, Cao P (2015) Preliminary study of improved SMR method based on gray system theory. Rock Soil Mech 36:408–412

Liu SF, Lu SF, Wan ZJ, Cheng JY (2019) Investigation of the influence mechanism of rock damage on rock fragmentation and cutting performance by the discrete element method. R Soc Open Sci 6(5):190116

Miao FS, Wu YP, Xie YH, Li YH (2018) Prediction of landslide displacement with step-like behavior based on multialgorithm optimization and a support vector regression model. Landslides 15:475–488

Mitchell A, Hungr O (2017) Theory and calibration of the Pierre 2 stochastic rock fall dynamics simulation program. Can Geotech J 54(1):18–30

Morse MS, Lu N, Wayllace A, Godt JW (2017) Evolution of Strain localization in variable-width three-dimensional unsaturated laboratory-scale cut slopes. J Eng Mech 143(9):04017085

Nilsen B (2017) Rock slope stability analysis according to Eurocode 7, discussion of some dilemmas with particular focus on limit equilibrium analysis. Bull Eng Geol Env 76(4):1229–1236

Qin ZC, Li T, Li QH, Chen GB, Cao B (2019) Mechanism of rock burst based on energy dissipation theory and its applications in erosion zone. Acta Geodynamica Et Geromaterialia 16(2):119–130

Sarfaraz H, Amini M (2020) Numerical Modeling of rock slopes with a potential of block-flexural toppling failure. J Min Environ 11(1):247–259

Scaringi G, Fan XM, Xu Q, Liu C, Ouyang CJ, Domenech G, Yang F, Dai LX (2018) Some considerations on the use of numerical methods to simulate past landslides and possible new failures: the case of the recent Xinmo landslide (Sichuan, China). Landslides 15(7):1359–1375

Shen H, Klapperich H, Abbas SM, Ibrahim A (2012) Slope stability analysis based on the integration of GIS and numerical simulation. Autom Constr 26:46–53

Shi GH (1988) Discontinuous deformation analysis: a new numerical model for the statics and dynamics of block systems. University of California, Berkeley

Simeoni L, Mongiovì L (2007) Inclinometer monitoring of the Castelrotto landslide in Italy. J Geotech Geoenviron Eng 133(6):653–666

Stead D, Wolter A (2015) A critical review of rock slope failure mechanisms: the importance of structural geology. J Struct Geol 74:1–23

Stead D, Eberhardt E, Coggan JS (2006) Developments in the characterization of complex rock slope deformation and failure using numerical modelling techniques. Eng Geol 83(1–3):217–235
Sun SR, Xu PL, Wu JM, Wei JH, Fu WG, Liu J, Debi PK (2014) Strength parameter identification and application of soil-rock mixture for steep-walled talus slopes in southwestern China. Bull Eng Geol Env 73(1):123–140
Sun P, Wang G, Wu LZ, Igwe O, Zhu EZ (2019) Physical model experiments for shallow failure in rainfall-triggered loess slope, Northwest China. Bull Eng Geol Env 78(6):4363–4382
Tang CA (1997) Numerical simulation of progressive rock failure and associated seismicity. Int J Rock Mech Min Sci 34(2):249–250
Tomas R, Cuenca A, Cano M, García-Barba J (2012) A graphical approach for slope mass rating (SMR). Eng Geol 124:67–76
Tonon F (2020) Simplified consideration for permanent rock dowels in block theory and 2-D limit equilibrium analyses. Rock Mech Rock Eng 53(4):2001–2006
Wang LQ, Yin YP, Huang BL, Dai ZW (2020a) Damage evolution and stability analysis of the Jianchuandong dangerous rock mass in the three gorges reservoir area. Eng Geol 265:105439
Wang LQ, Yin YP, Huang BL, Zhang ZH, Zhao P, Wei YJ (2020b) A study of the treatment of a dangerous thick submerged rock mass in the three gorges reservoir area. Bull Eng Geol Env 79(5):2579–2590
Wang LQ, Huang BL, Zhang ZH, Dai ZW, Zhao P, Hu MJ (2020c) The analysis of slippage failure of the HuangNanBei slope under dry-wet cycles in the three gorges reservoir region, China. Geomat Nat Haz Risk 11(1):1233–1249
Wang Z, Liu H, Gao X, Thomas B, Zhang W (2020f) Stability analysis of soil slopes based on strain information. Acta Geotech 15(11):3121–3134
Wang ZY, Gu DM, Zhang WG (2020e) Influence of excavation schemes on slope stability: a DEM study. J Mountain Sci 17(6):1509–1522
Wu LZ, Huang RQ, Xu Q, Zhang LM, Li HL (2015) Analysis of physical testing of rainfall-induced soil slope failures. Environ Earth Sci 73(12):8519–8531
Xia KZ, Liu XM, Chen CX, Song YF, Ou Z, Long Y (2015) Analysis of mechanism of bedding rock slope instability with catastrophe theory. Rock Soil Mech 32(2):477–486
Xing AG, Yuan XY, Xu Q, Zhao QH, Huang HQ, Cheng QG (2017) Characteristics and numerical runout modelling of a catastrophic rock avalanche triggered by the Wenchuan earthquake in the Wenjia valley, Mianzhu, Sichuan, China. Landslides 14(1):83–98
Xu Q, Peng D, Zhang S, Zhu X, He C, Qi X, Zhao K, Xiu D, Ju N (2020) Successful implementations of a real-time and intelligent early warning system for loess landslides on the Heifangtai terrace, China. Eng Geol 278:105817
Yu YJ, Zou CN, Dong DZ, Wang SJ, Li JZ, Yang H, Li DH, Li XJ, Wang YM, Huang JL (2014) Geological Conditions and prospect forecast of shale gas formation in Qiangtang Basin, Qinghai-Tibet Plateau. Acta Geol Sin-English Edition 88(2):598–619
Zhang W, Tang L, Li H, Wang L, Cheng L, Zhou T, Chen X (2020) Probabilistic Stability analysis of bazimen landslide with monitored rainfall data and water level fluctuations in TGR, China. Front Struct Civ Eng 14(5):1247–1261
Zhang WG, Meng FS, Chen FY, Liu HL (2021) Effects of spatial variability of weak layer and seismic randomness on rock slope stability and reliability analysis. Soil Dyn Earthq Eng 146:106735. https://doi.org/10.1016/j.soildyn.2021.106735
Zhou X, Chen JP, Chen Y, Song SY, Shi MY, Zhan JW (2017) Bayesian-based probabilistic kinematic analysis of discontinuity-controlled rock slope instabilities. Bull Eng Geol Env 76(4):1249–1262
Zhu DY, Lee CF, Jiang HD (2003) Generalized framework of limit equilibrium methods for slope stability analysis. Geotechnique 53(4):377–395
Zhu X, Xu Q, Zhou J, Deng M (2012) Remote Landslide observation system with differential GPS. Procedia Earth Planet Sci 5:70–75
Zhu X, Xu Q, Tang M, Nie W, Ma S, Xu Z (2017b) Comparison of two optimized machine learning models for predicting displacement of rainfall-induced landslide: a case study in Sichuan Province, China. Eng Geol 218:213–222

Zhu X, Xu Q, Tang M, Li H, Liu F (2018) A hybrid machine learning and computing model for forecasting displacement of multifactor-induced landslides. Neural Comput Appl 30(12):3825–3835

Zhu X, Xu Q, Qi X, Liu H (2017a) A self-adaptive data acquistion technique and its application in landslide monitoring. In: Mikoš MA, Arbanas Ž, Yin Y, Sassa K (eds) Proceedings of 4th world landslide forum, pp 71–78

Chapter 2
Machine Learning Algorithms

2.1 Supervised Learning

Supervised learning, also known as supervised machine learning, is defined by its use of labeled datasets to train algorithms that to classify data or predict outcomes accurately. As input data is fed into the model, it adjusts its weights until the model has been fitted appropriately. This occurs as part of the cross-validation process to ensure that the model avoids overfitting or underfitting. Supervised learning helps organizations solve for a variety of real-world problems at scale, such as classifying spam in a separate folder from your inbox. Some methods used in supervised learning include logistic regression, back propagation neural network (Goh 1995), etc. Figure 2.1 provides the VOSviewer plots for some main supervised learning methods, in which the keyword co-occurrence network of supervised learning was constructed by the VOSviewer software. The size of the nodes and words represents the weights of the nodes. The bigger the node and word are, the larger the weight is. The distance between two nodes reflects the strength of the relation between two nodes. A shorter distance generally reveals a stronger relationship. The line between two keywords represents that they have appeared together. The thicker the line is, the more co-occurrence they have. The nodes with the same color belong to a cluster.

2.2 Unsupervised Learning

Unsupervised learning, also known as unsupervised machine learning, uses machine learning algorithms to analyze and cluster unlabeled datasets. These algorithms discover hidden patterns or data groupings without the need for human intervention. Its ability to discover similarities and differences in information makes it the ideal solution for exploratory data analysis, cross-selling strategies, customer segmentation, image and pattern recognition. It's also used to reduce the number of features in a model through the process of dimensionality reduction; k-means clustering (Qin

Z. Wengang et al., *Application of Machine Learning in Slope Stability Assessment*,
https://doi.org/10.1007/978-981-99-2756-2_2

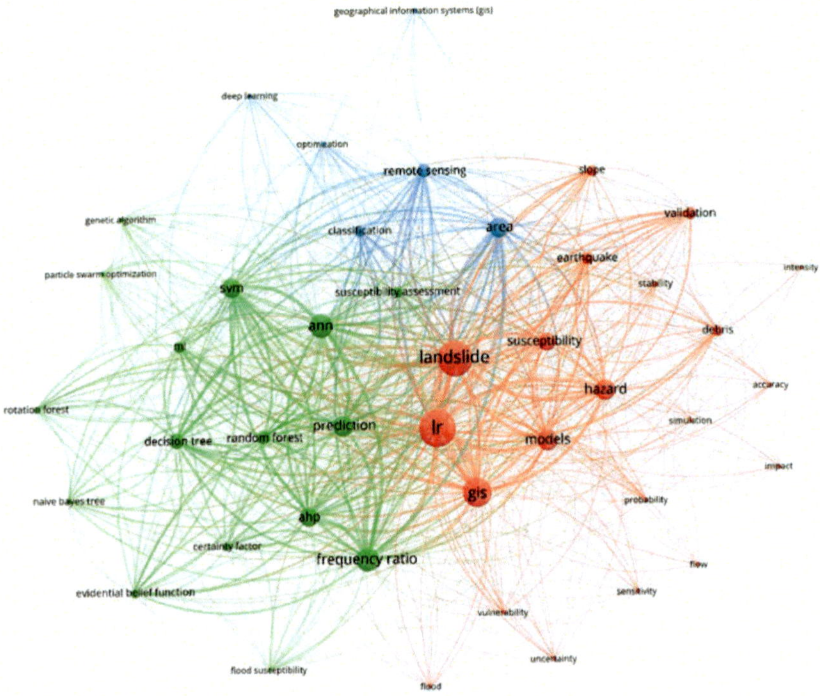

Fig. 2.1 VOSviewer plots for some main supervised learning methods

et al. 2018), Apriori (Miao et al. 2021), and singular value decomposition are three common approaches for this, as shown in Fig. 2.2.

2.3 Semi-supervised Learning

Semi-supervised learning offers a happy medium between supervised and unsupervised learning. During training, it uses a smaller labeled dataset to guide classification and feature extraction from a larger, unlabeled dataset. Semi-supervised learning can solve the problem of having not enough labeled data (or not being able to afford to label enough data) to train a supervised learning algorithm. The main semi-supervised learning algorithms include: Graph inference and Laplacian support vector machine (SVM). Figure 2.3 provides the VOSviewer plots for some main semi-supervised learning methods.

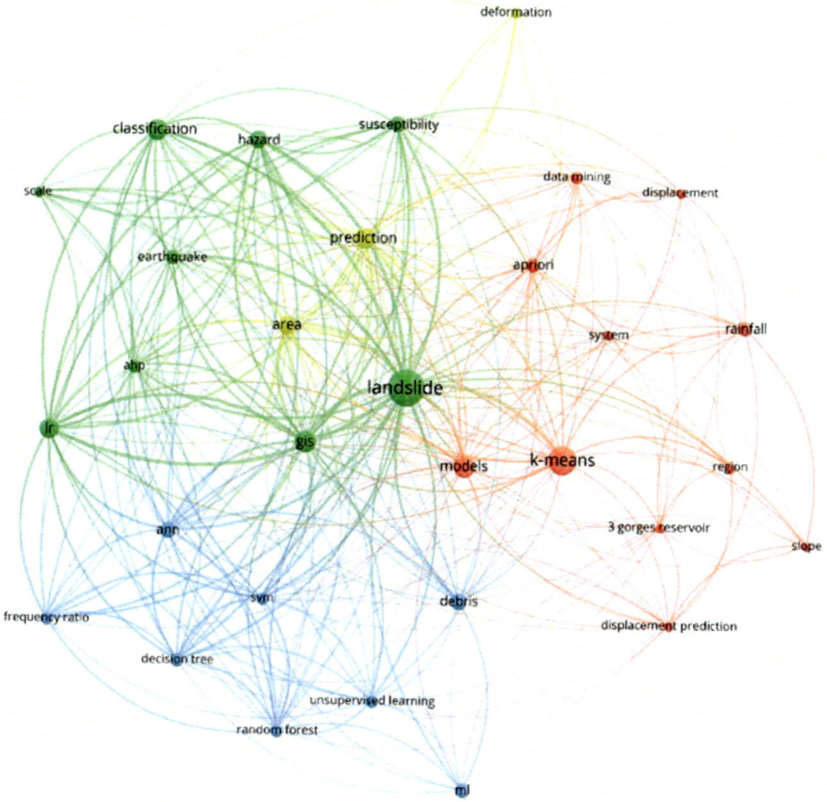

Fig. 2.2 VOSviewer plots for some main unsupervised learning methods

2.4 Reinforcement Learning

Reinforcement learning (RL) is a branch of machine learning that is based on actions, states, and rewards. A reinforcement learning agent is given a set of actions that it can apply to its environment to obtain rewards or reach a certain goal. These actions create changes to the state of the agent and the environment. The RL agent receives rewards based on how its actions bring it closer to its goal. RL agents usually start by knowing nothing about their environment and selecting random actions. As they gradually receive feedback from their environment, they learn sequences of actions that can maximize their rewards. Reinforcement learning has also helped researchers master complicated games such as Go, StarCraft 2, and DOTA. The use of RL in slope engineering and landslide is limited, mainly comprising of the Q-Learning and temporal difference learning. Figure 2.4 depicts the VOSviewer plots for main RL methods.

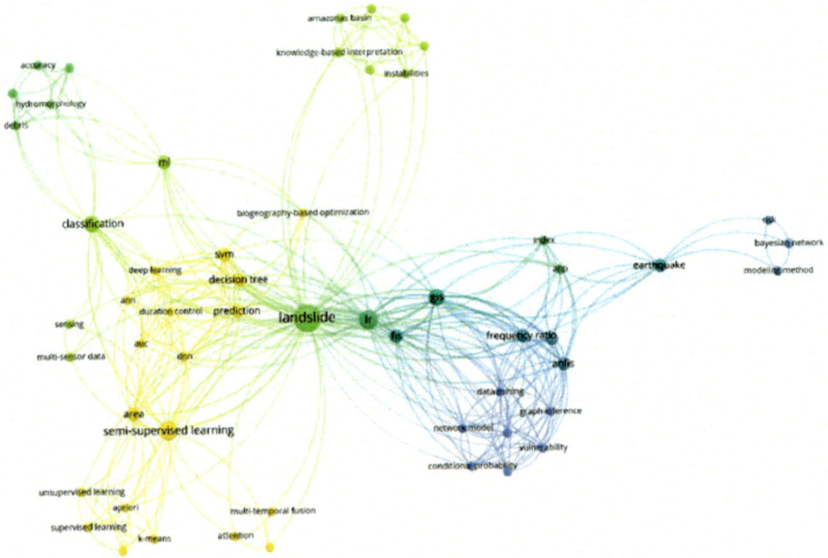

Fig. 2.3 VOSviewer plots for some main semi-supervised learning methods

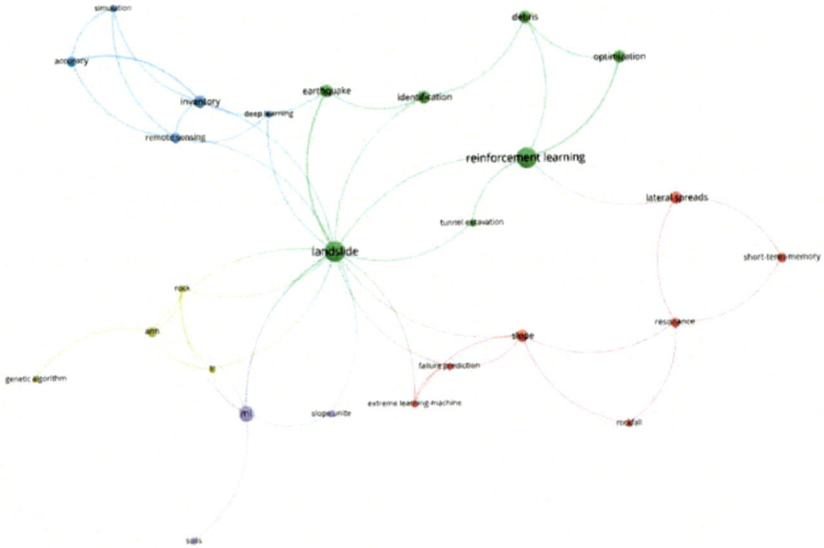

Fig. 2.4 VOSviewer plots for main RL methods

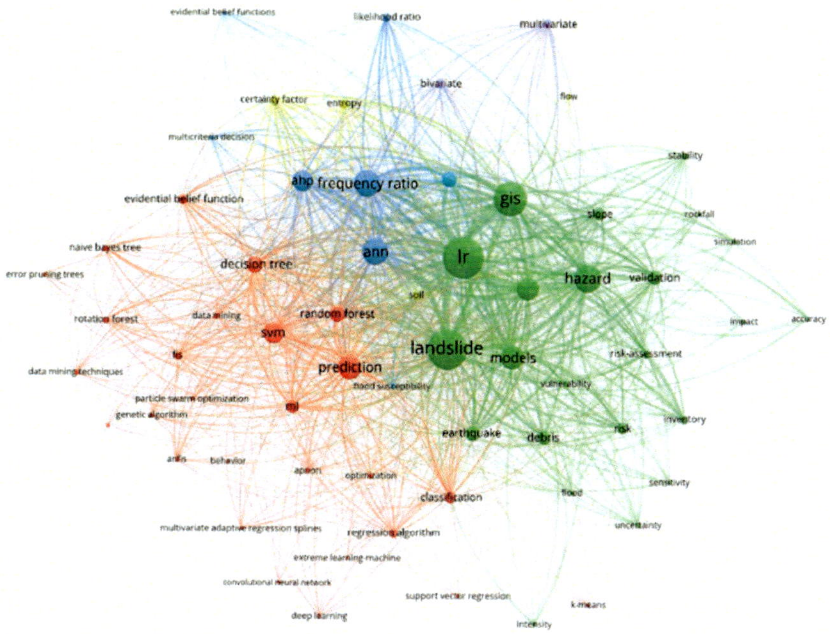

Fig. 2.5 VOSviewer plots for main regression algorithms

2.5 Regression Algorithm

Regression is another important and broadly used statistical and machine learning tool. The key objective of regression-based tasks is to predict output labels or responses which are continues numeric values, for the given input data. The output will be based on what the model has learned in training phase. Basically, regression models use the input data features (independent variables) and their corresponding continuous numeric output values (dependent or outcome variables) to learn specific association between inputs and corresponding outputs. The main regression algorithms include: Ordinary Least Square, Logistic Regression, Stepwise Regression, Multivariate Adaptive Regression Splines, Locally Estimated Scatterplot Smoothing, etc. Figure 2.5 shows the VOSviewer plots for main regression algorithms.

2.6 Case-Based Algorithm

Case-based reasoning (CBR) is an expert system development methodology which reuses past solutions to solve new problems. The first approach uses CBR to retain K similar cases to solve the motion planning problem by merging those

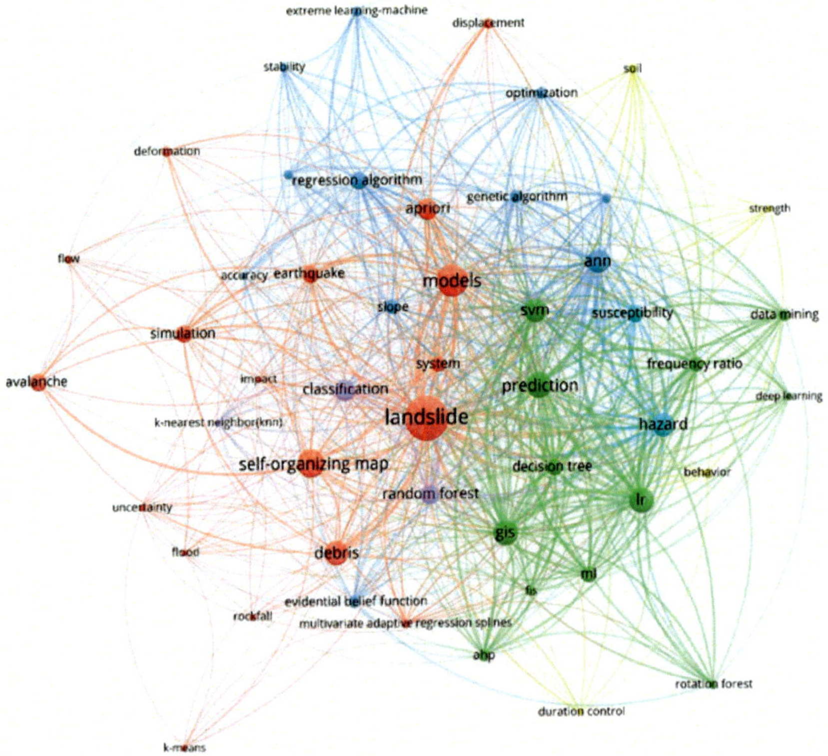

Fig. 2.6 VOSviewer plots for main CBR algorithms

solutions into a set. Afterward, it picks from this set based on a heuristic function to assemble a final solution. Regarding the second approach, it employs the retained K similar cases differently. It uses those solution to build a graph which can be queried using traditional graph search algorithms. Results prove the success of such approaches concerning solution quality and success rate compared with different experience-based algorithms. The main CBR algorithms include the K-nearest neighbor, learning vector quantization, and self-organizing map (SOM). Figure 2.6 shows the VOSviewer plots for main CBR algorithms.

2.7 Regularization Method

Regularization is a technique which makes slight modifications to the learning algorithm such that the model generalizes better. This in turn improves the model's performance on the unseen data as well. How does regularization help reduce overfitting? Let's consider a neural network which is overfitting on the training data. If

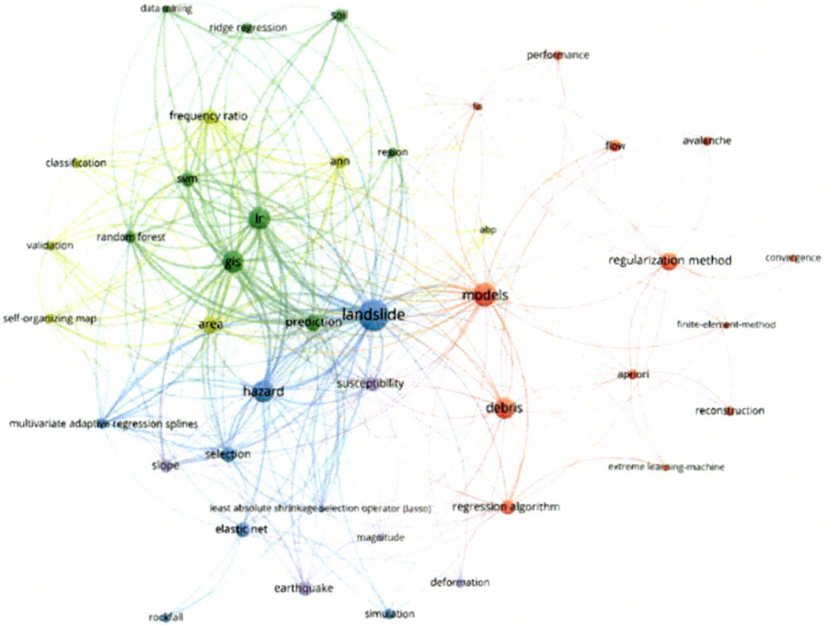

Fig. 2.7 VOSviewer plots for main regularization algorithms

you have studied the concept of regularization in machine learning, you will have a fair idea that regularization penalizes the coefficients. In deep learning, it actually penalizes the weight matrices of the nodes. Assume that our regularization coefficient is so high that some of the weight matrices are nearly equal to zero. This will result in a much simpler linear network and slight underfitting of the training data. Such a large value of the regularization coefficient is not that useful. We need to optimize the value of regularization coefficient in order to obtain a well-fitted model. Different regularization techniques include the ridge regression, least absolute shrinkage and selection operator, and elastic net, etc. Figure 2.7 shows the VOSviewer plots for main regularization algorithms.

2.8 Decision Tree

Decision trees (DTs) are a nonparametric supervised learning method used for classification and regression. The goal is to create a model that predicts the value of a target variable by learning simple decision rules inferred from the data features. A tree can be seen as a piecewise constant approximation, and a decision tree is a flowchart like tree structure, where each internal node denotes a test on an attribute, each branch represents an outcome of the test, and each leaf node (terminal node)

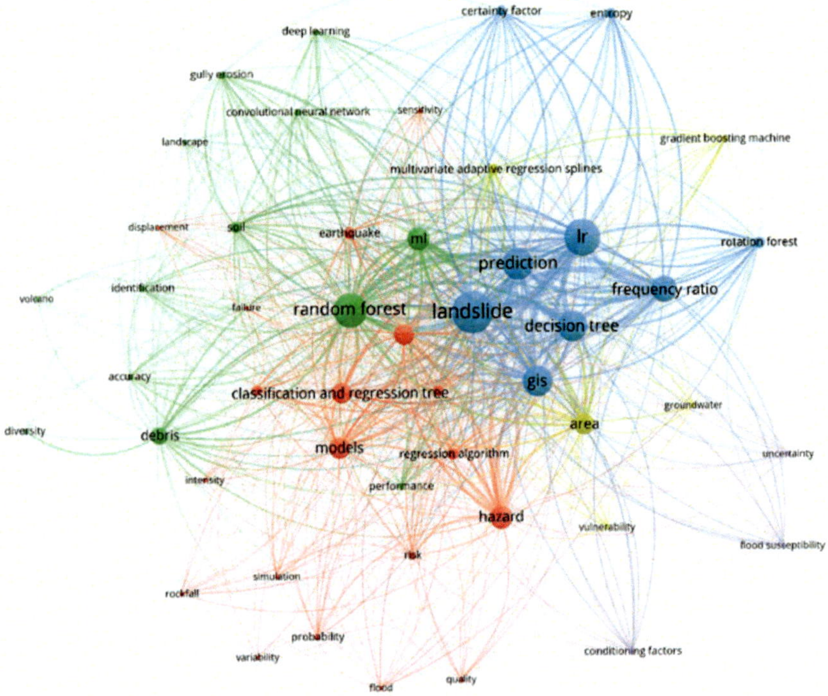

Fig. 2.8 VOSviewer plots for main DT algorithms

holds a class label. The main DT algorithms include the classification and regression tree, ID3 (Iterative Dichotomiser 3), C4.5, chi-squared automatic interaction detection, decision stump, random forest, and gradient boosting machine (GBM). Figure 2.8 shows the VOSviewer plots for main DT algorithms.

2.9 Bayesian Method

In recent years, Bayesian learning has been widely adopted and even proven to be more powerful than other machine learning techniques (Zhao et al. 2018; Zhao and Wang 2020). For example, we have seen that recent competition winners are using Bayesian learning to come up with state-of-the-art solutions to win certain machine learning challenges. This shows that Bayesian learning is the solution for cases where probabilistic modeling is more convenient and traditional machine learning

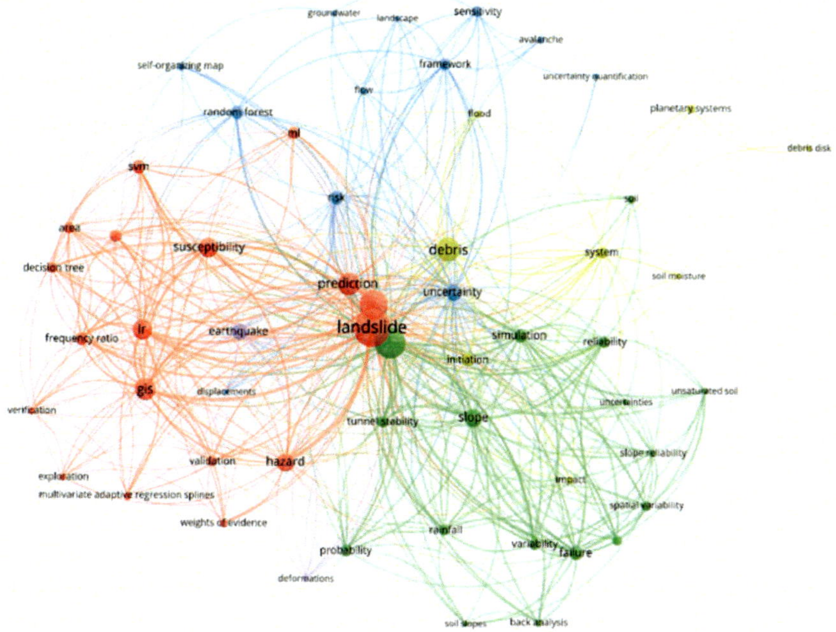

Fig. 2.9 VOSviewer plots for main Bayesian algorithms

techniques fail to provide state-of-the-art solutions. Bayesian-based approaches play a significant role in data science owing to the following unique capabilities:

- Incorporating prior knowledge or beliefs with the observed data to determine the final posterior probability
- New observations or evidence can incrementally improve the estimated posterior probability
- The ability to express uncertainty in predictions

The main Bayesian algorithms include the Naive Bayes, the averaged one-dependence estimators, and the Bayesian belief network. Figure 2.9 shows the VOSviewer plots for main Bayesian algorithms.

2.10 Kernel-Based Algorithm

In the last years, a number of powerful kernel-based learning machines, e.g., SVM, radial basis function, Kernel fisher discriminant, linear discriminate analysis, and Kernel principal component analysis, have been proposed. These approaches have shown practical relevance not only for classification problems but also for regression issues. Successful applications of kernel-based algorithms have been reported for

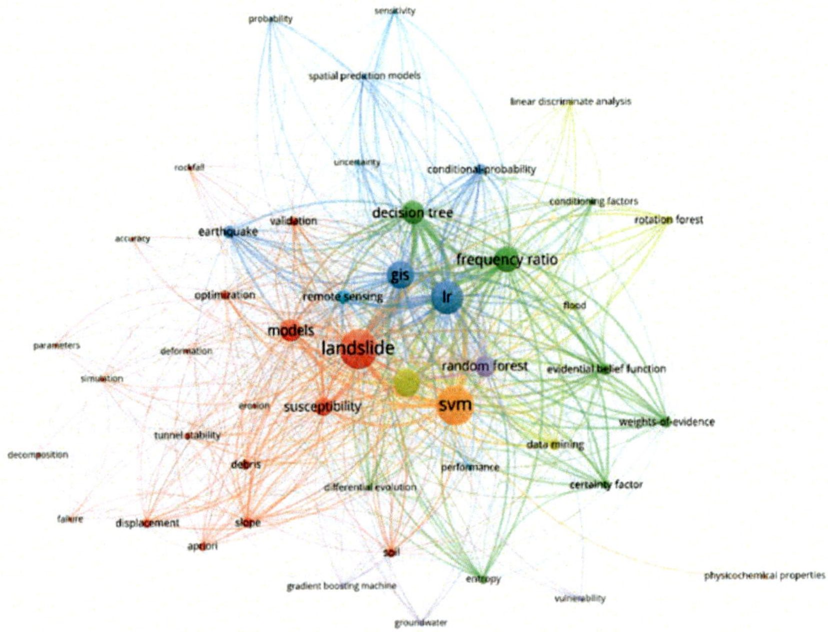

Fig. 2.10 VOSviewer plots for main kernel-based algorithms

various fields including slope engineering. Figure 2.10 shows the VOSviewer plots for main kernel-based algorithms.

2.11 Clustering

Clustering is a kind of unsupervised machine learning, where there will be only feature or input columns. In simple terms, there is no supervision given for the model by label or target column. Such models try to learn by grouping or segmentation of data points/ datasets. The data points forming the individual cluster should have similar features or properties. In simple words, each cluster contains the same property, not a different property. The different property data points are grouped to form another set of clusters.

Whenever a dataset is given, we try to find some meaningful insights from the given data point. In real-time, we make use of clustering methods in most of the problem statements. The most widely used popular clustering algorithms are: K-means clustering, hierarchical clustering, density-based spatial clustering of applications with noise. Figure 2.11 shows the VOSviewer plots for main clustering algorithms.

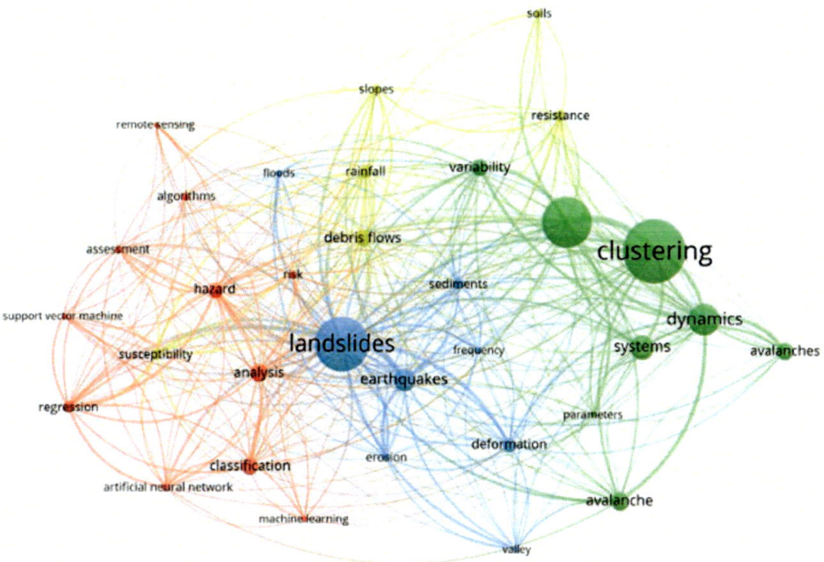

Fig. 2.11 VOSviewer plots for main clustering algorithms

2.12 Association Rule Learning

Association rule learning is a type of unsupervised learning technique that checks for the dependency of one data item on another data item and maps accordingly so that it can be more profitable. It tries to find some interesting relations or associations among the variables of dataset. It is based on different rules to discover the interesting relations between variables in the database. Association rule learning can be divided into three types of algorithms: Apriori, Eclat, and F-P growth algorithm. Figure 2.12 shows the VOSviewer plots for main association rule learnings.

2.13 Artificial Neural Network

Artificial neural network (ANN) intended to simulate the behavior of biological systems composed of "neurons". ANNs are computational models inspired by an animal's central nervous systems. It is capable of machine learning as well as pattern recognition. These presented as systems of interconnected "neurons" which can compute values from inputs. A neural network is an oriented graph. It consists of nodes which in the biological analogy represent neurons, connected by arcs. It corresponds to dendrites and synapses. Each arc associated with a weight while at each node. Apply the values received as input by the node and define activation function along the incoming arcs, adjusted by the weights of the arcs.

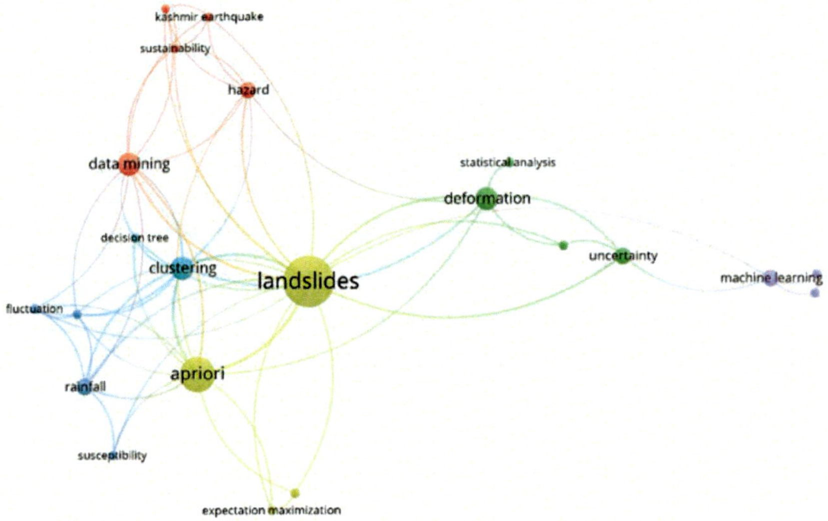

Fig. 2.12 VOSviewer plots for main association rule learnings

A neural network is a machine learning algorithm based on the model of a human neuron. The human brain consists of millions of neurons. It sends and process signals in the form of electrical and chemical signals. These neurons are connected with a special structure known as synapses. Synapses allow neurons to pass signals. From large numbers of simulated neurons, neural networks form. ANN can be divided into the following main types of algorithms: perceptron neural network, back propagation neural networks, Hopfield neural networks, SOM, etc. Figure 2.13 shows the VOSviewer plots for main ANNs.

2.14 Deep Learning

Deep learning is a sub-field of machine learning concerned with algorithms inspired by the structure and function of the brain called artificial neural networks. It is large neural networks with huge amounts of data to be proceeded (great scale). In addition to scalability, another often cited benefit of deep learning models is their ability to perform automatic feature extraction from raw data, also called feature learning. Modern state-of-the-art deep learning focused on training deep (many layered) neural network models using the backpropagation algorithm and the most popular techniques are: multi-layer perceptron networks, CNN, long short-term memory recurrent neural networks, restricted Boltzmann machine, deep belief networks, stacked auto-encoders, etc. Figure 2.14 shows the VOSviewer plots for main deep learning algorithms.

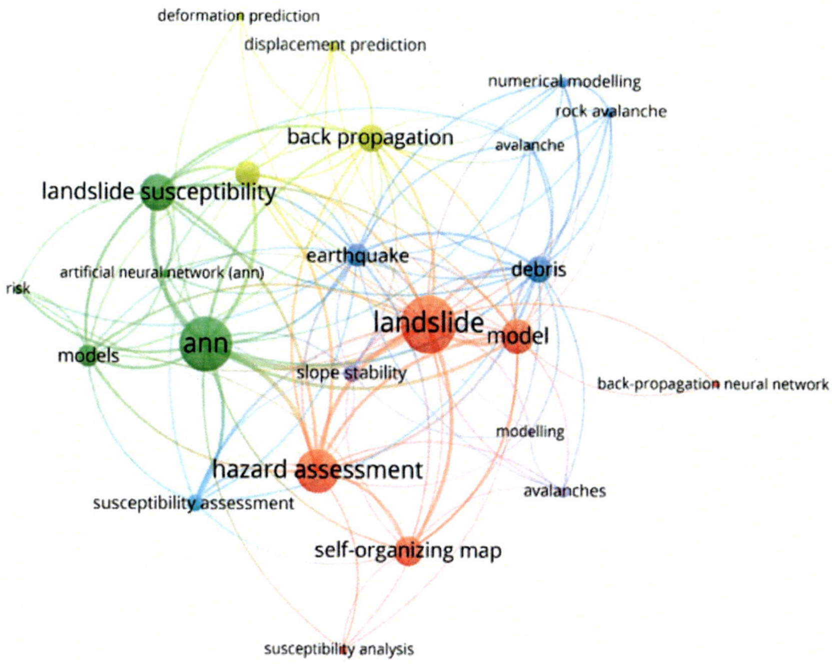

Fig. 2.13 VOSviewer plots for main ANNs

2.15 Dimension Reduction

Dimensionality reduction is the process of reducing the number of random variables under consideration, by obtaining a set of principal variables. It can be divided into feature selection and feature extraction. An intuitive example of dimensionality reduction can be discussed through a simple e-mail classification problem, where we need to classify whether the e-mail is spam or not. This can involve a large number of features, such as whether or not the e-mail has a generic title, the content of the e-mail, whether the e-mail uses a template, etc. However, some of these features may overlap. In another condition, a classification problem that relies on both humidity and rainfall can be collapsed into just one underlying feature, since both of the aforementioned are correlated to a high degree. Hence, we can reduce the number of features in such problems. A 3-D classification problem can be hard to visualize, whereas a 2-D one can be mapped to a simple two dimensional space, and a 1-D problem to a simple line. Figure 2.15 illustrates this concept, where a 3-D feature space is split into two 1-D feature spaces, and later, if found to be correlated, the number of features can be reduced even further.

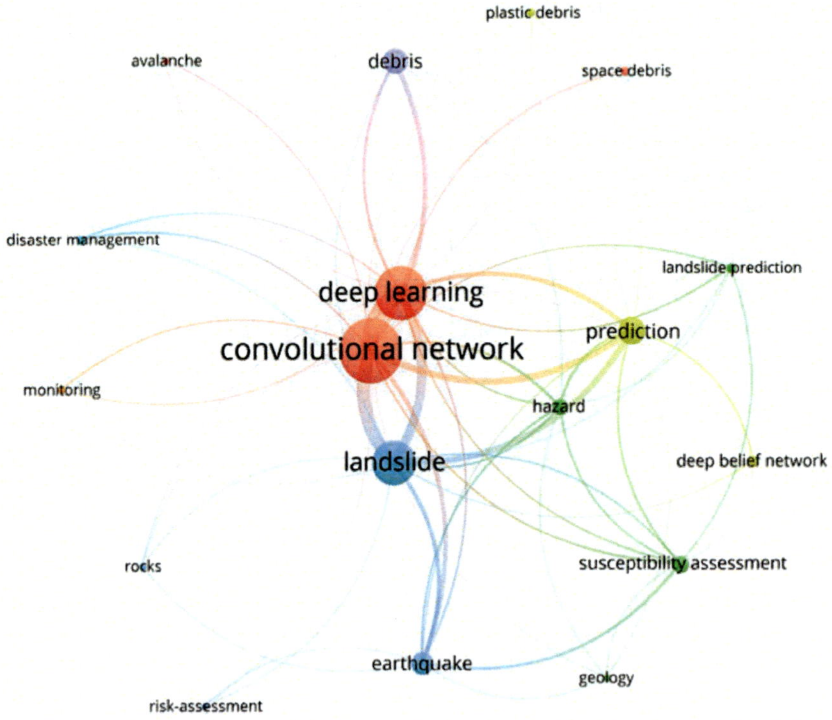

Fig. 2.14 VOSviewer plots for main deep learning algorithms

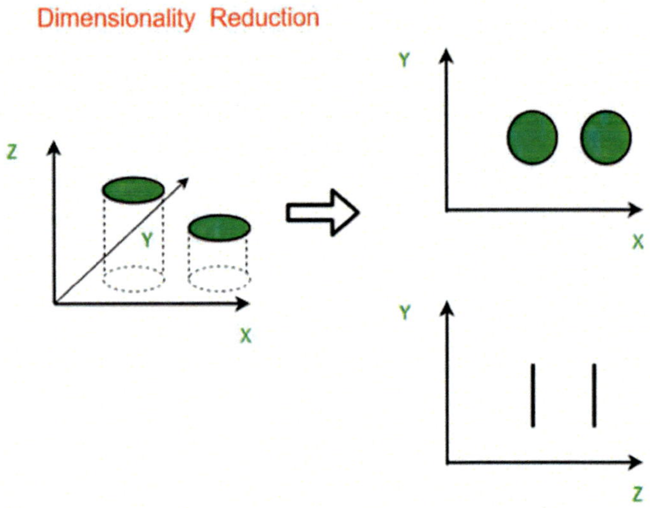

Fig. 2.15 Illustration of dimension reduction

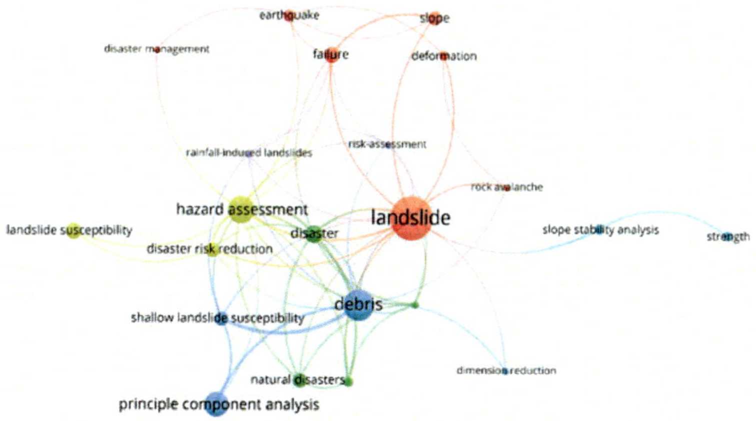

Fig. 2.16 VOSviewer plots for main dimension reduction techniques

The various methods used for dimensionality reduction include: principal component analysis, linear discriminant analysis, generalized discriminant analysis, partial least square regression, multi-dimensional scaling, and projection pursuit. Figure 2.16 shows the VOSviewer plots for main dimension reduction techniques.

2.16 Ensemble Learning

Ensemble learning is the process by which multiple models (e.g., Dong et al. 2020; Sahin 2020; Wang et al. 2020a, b; Zhang et al. 2021a, b), such as classifiers or experts, are strategically generated and combined to solve a particular computational intelligence problem. Ensemble learning is primarily used to improve the (classification, prediction, function approximation, etc.) performance of a model, or reduce the likelihood of an unfortunate selection of a poor one. Other applications of ensemble learning include assigning a confidence to the decision made by the model, selecting optimal (or near optimal) features, data fusion, incremental learning, non-stationary learning, and error-correcting. The various methods used for ensemble learning include: boosting, bootstrapped aggregation (Bagging), AdaBoost, stacked generalization, GBM, and random forest (RF). Figure 2.17 shows the VOSviewer plots for main ensemble learning algorithms.

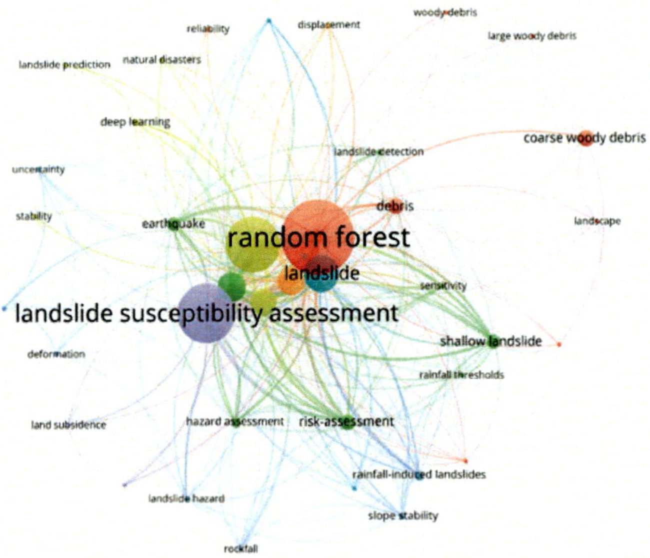

Fig. 2.17 VOSviewer plots for main ensemble learning algorithms

References

Dong W, Huang Y, Lehane B, Ma G (2020) XGBoost algorithm-based prediction of concrete electrical resistivity for structural health monitoring. Autom Constr 114:103155. https://doi.org/10.1016/j.autcon.2020.103155

Goh ATC (1995) Back-propagation neural networks for modeling complex systems. Artif Intell Eng 9:143–151. https://doi.org/10.1016/0954-1810(94)00011-S

Miao F, Wu Y, Li L, Liao K, Xue Y (2021) Triggering factors and threshold analysis of Baishuihe landslide based on the data mining methods. Nat Hazards 105(3):2677–2696. https://doi.org/10.1007/s11069-020-04419-5

Qin YQ, Tang H, Feng ZY, Yin XT, Wang DY (2018) Slope stability evaluation by clustering analysis. Yantu Lixue/Rock Soil Mech 39(8). https://doi.org/10.16285/j.rsm.2016.2756

Sahin EK (2020) Comparative analysis of gradient boosting algorithms for landslide susceptibility mapping. Geocarto Int 0(0):1–25. https://doi.org/10.1080/10106049.2020.1831623

Wang L, Wu C, Tang L, Zhang W, Lacasse S, Liu H, Gao L (2020a) Efficient reliability analysis of earth dam slope stability using extreme gradient boosting method. Acta Geotech 15(11):3135–3150. https://doi.org/10.1007/s11440-020-00962-4

Wang MX, Huang DR, Wang G, Li DQ (2020b) SS-XGBoost: a machine learning framework for predicting newmark sliding displacements of slopes. J Geotech Geoenviron Eng 146(9):04020074. https://doi.org/10.1061/(asce)gt.1943-5606.0002297

Zhang W, Wu C, Zhong H, Li Y, Wang L (2021a) Prediction of undrained shear strength using extreme gradient boosting and random forest based on Bayesian optimization. Geosci Front 12(1):469–477. https://doi.org/10.1016/j.gsf.2020.03.007

Zhang W, Wu C, Li Y, Wang L, Samui P (2021b) Assessment of pile drivability using random forest regression and multivariate adaptive regression splines. Georisk Assess Manage Risk Eng Syst Geohazards 15(1):27–40. https://doi.org/10.1080/17499518.2019.1674340

Zhao T, Hu Y, Wang Y (2018) Statistical interpretation of spatially varying 2D geo-data from sparse measurements using Bayesian compressive sampling. Eng Geol 246(September):162–175. https://doi.org/10.1016/j.enggeo.2018.09.022

Zhao T, Wang Y (2020) Interpolation and stratification of multilayer soil property profile from sparse measurements using machine learning methods. Eng Geol 265(October 2019):105430. https://doi.org/10.1016/j.enggeo.2019.105430

Chapter 3
Real-Time Monitoring and Early Warning of Landslide

3.1 Introduction

Due to the climate changes, human activities, and geological structures, landslides become a major natural hazard causing not only thousands of deaths and injuries but also significant damage to property and infrastructure very year around the world (Pecoraro et al. 2018; Intrieri et al. 2019; Petley 2012). Except for active protective structural works, an efficient monitoring and early warning system is regarded as another powerful tool for landslide risk mitigation and being increasingly applied worldwide (Pecoraro et al. 2018; Azzam et al. 2011; Madhusudhan et al. 2018). One of the main components of the early warning systems is temporal prediction of landslide, which can be determined as the possible time of the failure of unstable slope. Based on the efficient temporal prediction, taking necessary measures in advance will significantly reduce casualties and property damage. In fact, different monitoring parameters are used to predict landslide, depending on the triggering factors and mechanisms of landslide. Typically, rainfall and underground water level monitoring data are normally employed as important indicators for predicting rainfall/water induced landslide. However, the landslide deformation caused by the same rainfall and water level variation is different due to different rock/soil structure characteristics. So, as an indirect indicator of instability of landslide, the most common use of precipitation data is to derive rainfall/water level thresholds for rough early warning rather than precise forecasting of landslide, because they are prone to false or missed alarms (Intrieri et al. 2019).

Prediction of landslide is a global challenge due to the complex nature of landslide mechanisms, which has attracted numerous concerns around the world (Fan et al. 2019; Intrieri et al. 2019; Carlà et al. 2017; Xu et al. 2020; Ju et al. 2015). To overcome this challenge, our team have developed a real-time landslide early warning system (LEWS) of the State Key Laboratory and Geohazard Prevention and Geoenvironment Protection (SKLGP), in which more than hundreds of multiple types of landslides (includes rockslide, loess soil slides, shallow slope failures, etc.) in Southwest of China are monitored in real-time (Xu et al. 2009, 2011, 2020; Ju et al. 2015),

Z. Wengang et al., *Application of Machine Learning in Slope Stability Assessment*,
https://doi.org/10.1007/978-981-99-2756-2_3

as shown in Fig. 3.1. Meanwhile, we also carried out many physical model tests to verify the three-state creep deformation characteristics that firstly presented by Saito in 1960s (Saito 1969). Based on the monitoring data of large number of landslides and physical model tests, we found that the landslide must undergo a tertiary creep phase that leads to final rupture or failure. According to the Saito's classic interpretation, the deformation of slope before failure has three-stage deformation phases: (1) primary creep (initial deformation); (2) secondary creep (constant deformation); and (3) tertiary creep (acceleration deformation). Xu et al. (2011) further divided the tertiary creep into three sub-phases: initial acceleration phase, medium-term acceleration phase, and highly accelerated phase (or called imminent sliding phase). The observation that the displacement–time curve becomes almost vertical during the phase of tertiary creep can be adopted as pre-warning precursor of landslide failure (Xu et al. 2011). Therefore, regardless of the driving factors, displacement and its derivatives (velocity and acceleration) are directly related with the stability conditions of slope. With adequate knowledge of the general evolution laws of landslide under gravity-driven, the SKLGP team developed an updated real-time early warning system that includes wireless sensor network, cloud server, early warning model, and software platform (Ju et al. 2020; Xu et al. 2020). This system has successfully predicted more than 10 landslides since 2017 and saved hundreds of lives in China.

In this chapter, we will present the novel framework of landslide early warning system, which includes real-time monitoring technique, comprehensive early

Fig. 3.1 Locations of successfully predicted landslides by the LEWS of SKLGP

warning model, and a WebGIS-based platform. And then, one of successful early warning cases for an abrupt rocky landslide in Southwestern China will be presented in details.

3.2 Real-Time Monitoring Network

A real-time and stable monitoring technique plays very important role in landslide early warning system because the quality of observation data has strong influence on the analysis performance of the early warning model. With the rapid development of electronic and wireless communication techniques, modern monitoring approach can be implemented through deploying various miniature wireless sensors on-site in a cost-effective way. Figure 3.2 shows the framework of remote sensing network for landslide monitoring. All the sensors can be divided into two categories: deformation monitoring sensors and trigger factors monitoring sensors. The deformation monitoring sensor includes GNSS, extensometer (or crack gauge), robotic total station for ground surface deformation observation, in-place inclinometer for underground deformation monitoring, and rain gauge for rainfall intensity monitoring because rainfall is one of the most triggers of landslide. All original observations can be transmitted to remote cloud server through cellular phone network (3G/4G) or satellite. In some mountain areas, there are not covered by general cellular phone network, and we need to construct a local wireless network controlled by a gateway, which fill the gap between field sensors and the remote cloud server via satellite message exchange (e.g., short message service via Beidou satellite developed by China).

3.3 Intelligent Early Warning System

3.3.1 Early Warning Model and Alert Criteria

In addition to the proficient sensing technology and real-time monitoring approach, a precise warning model is the most key component of one LEWS. Based on the assumption of creep theory (Tavenas and Leroueil 1981), many researchers conducted empirical models to predict the time of failure through geometrical arguments (Intrieri et al. 2019; Saito 1969; Segalini et al. 2018). The main idea of those models is to derive the relationship from displacement–time curve to fit the typical creep curve and calculate the time of failure through the inverse velocity method (Zhou et al. 2020). Although the prediction of landslide failure time plays very important role in the landslide early warning system, the accuracy of this method has to be strongly affected by the empirical constant parameters and the noise in the observed displacement–time curve. The forecasted time of failure may experience strong fluctuations when the displacement with low signal-to-noise ratio. Therefore, the time of failure

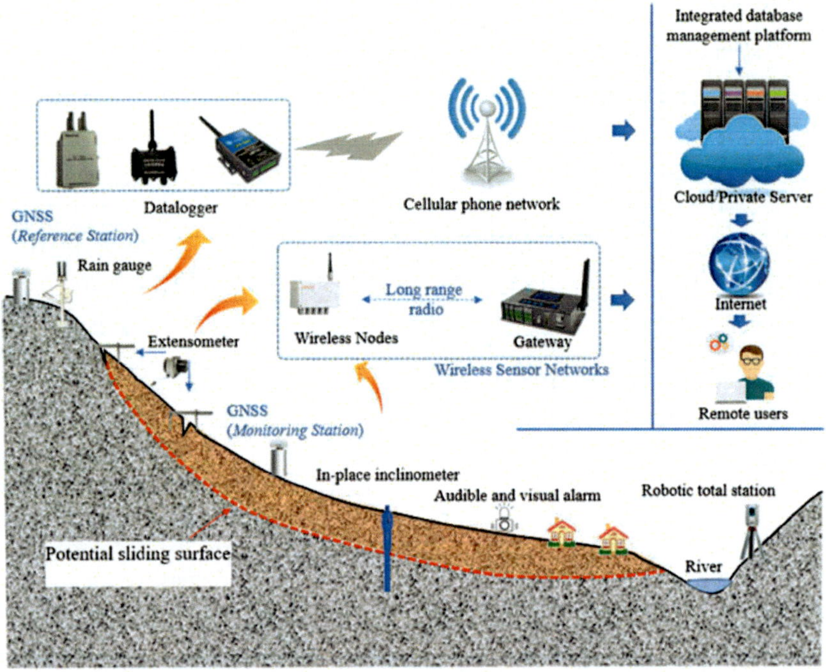

Fig. 3.2 Wireless sensor network for real-time landslide monitoring and early warning

method for landslide prediction is still under theoretical research, and few successful application cases have been reported.

Xu et al. (2011) developed an approach based on the observation that the displacement–time curve becomes almost vertical during the last phases of tertiary creep. To identify a general and quantitative criterion, the displacement is normalized by dividing it with respect to the average velocity of the secondary creep. In other words, the displacement–time (*S*-*t*) curve can be appropriately transformed through dividing the displacement (*S*) by average velocity (*v*) to produce a uniform dimensionally time plot in both axes (*T*-*t* curve shown in Fig. 3.3). The average velocity (*v*) is regarded as a constant value during the constant deformation. However, such value is difficult to obtain for rapid landslide because the constant deformation phase is very short. For this case, the long-term average velocity could be used to replace the velocity at the constant deformation phase. According to the *T*-*t* plot curve, tangential angle (α) referring to the deformation rate was presented to quantitatively and visually evaluate the slope *T*-*t* curve (Xu et al. 2011). The tangential angle (α) can be calculated as following:

$$\alpha = \arctan\frac{\Delta T}{\Delta t} = \arctan\frac{S_i - S_{i-1}}{\overline{v}(t_i - t_{i-1})} = \arctan\frac{v_i}{\overline{v}} \tag{3.4.1}$$

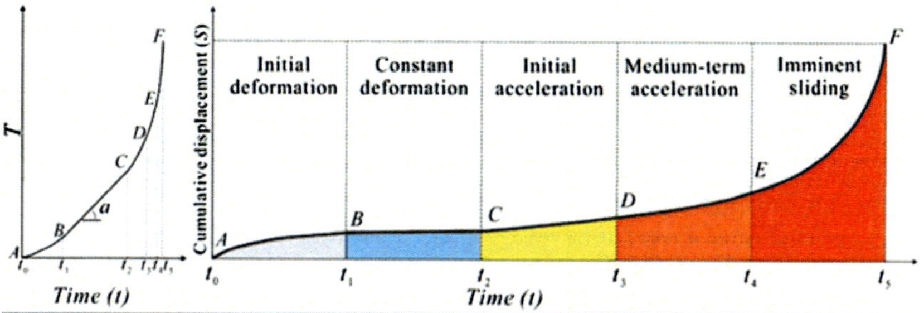

Corresponding deformation stage		Constant deformation	Acceleration deformation		
Alert parameters	Velocity V (mm/d)	$V_1 \leq V < V_2$	$V_2 \leq V < V_3$	$V \geq V_3$	
	Velocity increment ΔV (mm/d)	$\Delta V \approx 0$	$\Delta V > 0$		
	Tangential angle a (°)	$a \approx 45°$	$45° < a < 80°$	$80° \leq a < 85°$	$a \geq 85°$
Early warning level		Attention	Caution	Vigilance	Alarm

Note: V_1, V_2 and V_3 are the velocity threshold values in this comprehensive early warning model

Fig. 3.3 Outline of the four-level comprehensive warning for landslide based on deformation observation

$$v_i = \frac{S_i - S_{i-1}}{t_i - t_{i-1}} \tag{3.4.2}$$

$$T_i = \frac{S_i}{\overline{v}} \tag{3.4.3}$$

$$\overline{v} = \frac{S_i - S_0}{t_i - t_0} \tag{3.4.4}$$

where v_i is the daily average velocity; $\overline{v}$ is the overall average velocity; t_i is the current monitoring time period from the first observations.

The threshold that was proposed by Xu et al. (2011) is based on the tangential angle of the curve of *T*-*t* plot curve. As shown in Fig. 3.3, when the tangential angle is larger than 45°, the slope enters the tertiary creep stage; when the tangential angle is larger than 80°, the slope enters the medium acceleration phase; finally, if the angle is larger than 85°, the slope deformation enters the highly acceleration phase that corresponds to the imminent sliding condition.

Though the tangential angle method is unique and advantageous, defining a single threshold for all landslides will not be sufficient to make an precise judgment on the trend of the deformation (Fan et al. 2019). In practice, with aim to improve the robust and efficient of the early warning model, the overall average velocity ($\overline{v}$) and velocity increment (Δv) were combined with the tangential angle (α) which need to be considered as the comprehensive alert parameters for landslide prediction. The velocity increment (Δv) is used to identify the accelerated phases of deformation. The Δv does not change much and assumed to be around 0 in the constant deformation. At the beginning of the acceleration phase, the velocity increment (Δv) is mostly positive and larger than that of the constant deformation phase. The great growth of Δv usually indicates the premise of failure. Different landslides may exhibit different Δv. The velocity threshold based on statistical analyzes derived from monitoring data of landslides occurred in the past can be used to distinguish the true deformation of landslides from instrumental noise/errors. As shown in Fig. 3.3, V_1 mainly identifies the state when the slope begins to deform abnormally, while V_2 indicates whether the abnormal deformation enters a relatively fast degree, and V_3 indicates whether the deformation exceeds the short-time rapid deformation of the landslide. The multiple threshold method makes our early warning model more effective and robust.

3.3.2 3D-Web Early Warning System

In order to increase the reliability and intelligence of the LEWS, a computer-aid automatic operating system software was designed and deployed on a cloud server. Figure 3.4 shows the main interface of the 3D WebGIS-based landslide early warning system that was developed by the team of SKLGP. In the system, the field monitoring data (e.g., displacement, rainfall intensity, vibration, groundwater level, etc.) are collected in real-time via standard transmission protocol from the remote wireless sensor unit. And then, the data preprocessing and the comprehensive early warning algorithms are running in the server as a service program. All collected data is processed and analyzed in real-time. The analysis results and the original observed data are stored in database of cloud server and shown in the Web software simultaneously.

However, LEWS is not only a technology problem, but also a comprehensive societal issue. Missing and/or false warning may cause serious resident panic and loss of confidence in LEWS. Therefore, as shown in Fig. 3.5, a warning flowchart is designed to fill the gap between results of LEWS analysis and decision-making. To avoid false alarms, expert judgments should be a necessary work before final decision. When the alert parameters exceed the default alarm thresholds, pre-set warning information will be send to corresponding users through short message service, WeChat, Email, and other possible ways. When the yellow or red alert level is presented, an emergency online meeting will be informed to decide whether to change the alarm signal back to blue or to continue to issue a yellow or red alarm to the public in a formal way. Finally, after the alarm is issued, emergency response measures taken by the local government and the public are very import step to mitigate the landslide risk.

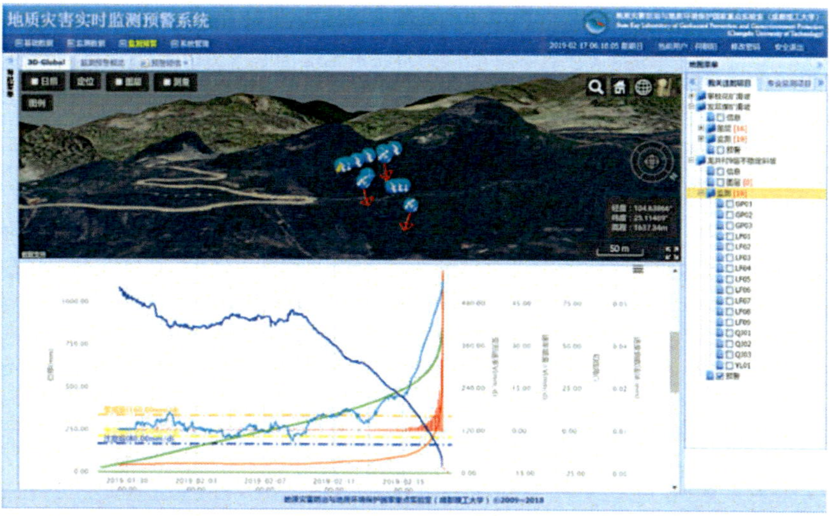

Fig. 3.4 WebGIS-based landslide early warning system platform

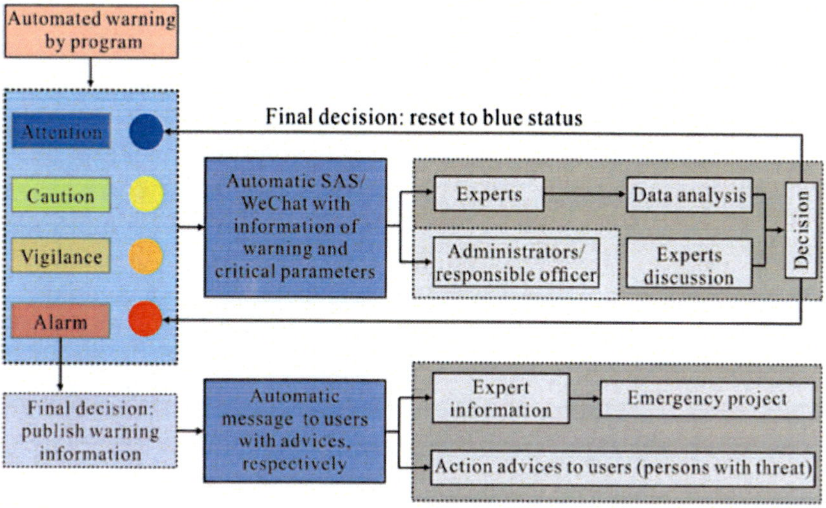

Fig. 3.5 Warning flowchart of LEWS (Xu et al. 2020; Ju et al. 2015)

3.4 Application

3.4.1 Introduction of Longjing Rocky Landslide

On 17 Feb. 2019, a large rocky landslide occurred in Longjing village, Xingyi City, Guizhou Province, China (Fan et al. 2019). The rocky landslide was identified about an estimated volume of 1.4 million m^3 that directly threatening the safety of more than 400 residents and one traffic road near the slope toe. To avoid property damage and casualties, emergency mitigation measures were implemented, and the above introduced LEWS of SKLGP was deployed on 27 Jan. 2019. Once found the deformation was entering the accelerating period, residents were evacuated, and the road was closed weeks before the final rockslide. A successful prediction of the rockslide was achieved 53 min before its occurrence at 05:53:17am on 17 Feb. 2019. Due to the successful early warning in advance, there was no casualties and property losses in this event. Figure 3.6 shows the aerial view of the Longjing rockslide. We found that the key to this successful LEWS mainly includes (i) the understanding of local geology and deformation history; (ii) the accurate position of monitoring sensors for landslide deformation; (iii) the comprehensive early warning model; (iv) immediate action to evacuate local residents; (v) and the responsible collaboration between local authorities and geological experts.

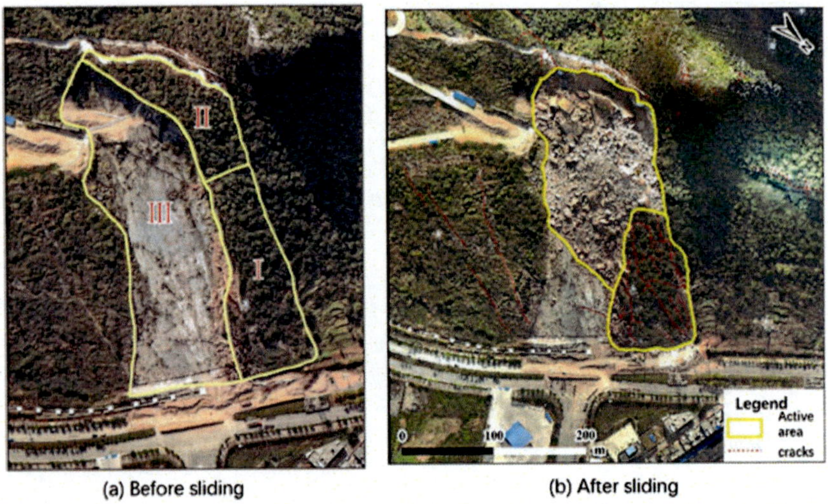

Fig. 3.6 Aerial view of Longjing Rockslide **a** before and **b** after failure

3.4.2 Geological Setting and Deformation History

As shown in Fig. 3.6a, the rockslide can be divided into I, II, III sub-zones according to the deformation characteristics and mechanism. Zone I and II are the main sliding rock mass of this landslide, and Zone III is the sliding surface of the previous landslide initially occurred in 2014 due to the removal of rock mass during the rock construction. Zone I mainly move along the sliding surface where a 2–5 cm thick interlayer of clay was found along the front edge of the rockslide. The relative height difference between the rockslide scarp and the road is around 252 m. Meanwhile, the slope angle of the potential sliding surface is about 45°. Consequently, the weak clay interlayer might make the rock mass prone to sliding under gravitational effect.

Figure 3.7 shows the deformation history of the rockslide in terms of image interpretation from 2014 to 2019. The result indicates how the evolution of this rockslide under the influence of human activities (e.g., road construction). In 2014, there was road construction and expansion of the existing road. The rock mass at the toe of slope has been excavated, which triggered the formation of the main scarp in orange. Once the road construction started, the deformation was accelerated due to further removal of the rock mass from the toe of slope. From the image captured in 2016, a large rockslide marked with yellow line in Fig. 3.7c should occurred from 2014 to 2016. Clear bedrock surface that shown in Fig. 3.7c indicated that the sliding rock mass has been completely removed to avoid further sliding triggered by rainfall. The trailing edge of the 2014 rockslide produced a free surface with a vertical height of about 25 m (Fig. 3.7d, e), which provide a favorable condition for the 17 Feb. 2019 rockslide.

There was no obvious deformation observed from the images after August 2016. In June 2018, a tensile crack with 0.2–0.9 m wide was observed at the crown of landslide. It developed into a major discontinuity with length of 200 m and width of 0.8–3.0 m by 5 December 2018. Finally, the rockslide occurred at 5:53:17am on 17 Feb. 2019, and the sliding boundary is shown in Fig. 3.7f.

3.4.3 Successful Monitoring and Early Warning

In this study, we mainly used displacement monitoring approaches because displacement and displacement-derived variables are the main alert parameters in the above-mentioned early warning model. The deployment of the various wireless sensor and its position on the site are indicated in Fig. 3.8. The crack gauges can be deployed along the main scarp to directly measure the displacement of rock mass. Meanwhile, the geodetic method GNSS is also deployed on the ground of landslide body. The difference between the crack gauge and GNSS is that the sampling frequency of crack gauge can reach to high frequency with high precision in real-time but GNSS just only reach to 30 min per sample. In this study, the crack gauge is a novel wireless end device that can adaptively adjust its sampling frequency with respect to the change

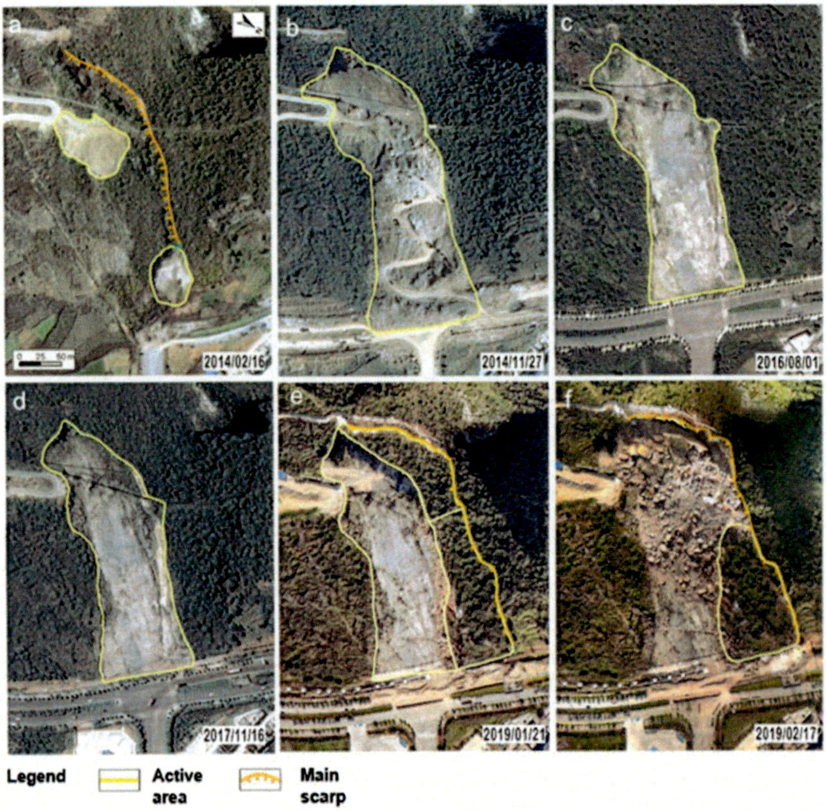

Fig. 3.7 Deformation history of the rockslide based on visual interpretation from Google Earth images (after Fan et al. 2019)

degree of landslide deformation (Zhu et al. 2017). Therefore, the self-adaptive smart crack gauge can capture the abrupt deformation of mass movement in time and obtain the high-quality displacement–time curve for more precise early warning of landslide. And the Tiltmeter was adopted to observe the deformation characteristics (e.g., topping or sliding) of Zone II (shown in Fig. 3.6) because there was a free vertical surface under it.

Figure 3.9 shows the overall monitoring process and calculated alert parameters based on the monitoring data by crack gauge in the developed WebGIS early warning system. The remote users can access this Web to know the evolution creep phases and weather it enters to different alert level. At the same time, once the alert parameters exceed the pre-set threshold value, the system can inform the administer and experts imminently. And Fig. 3.10 shows the detailed early warning information during acceleration phase that can be clearly divided into three sub-phases: Caution, Vigilance, and Alarm. At 5:00:00 am on 17 Feb. 2019, the system issued a red

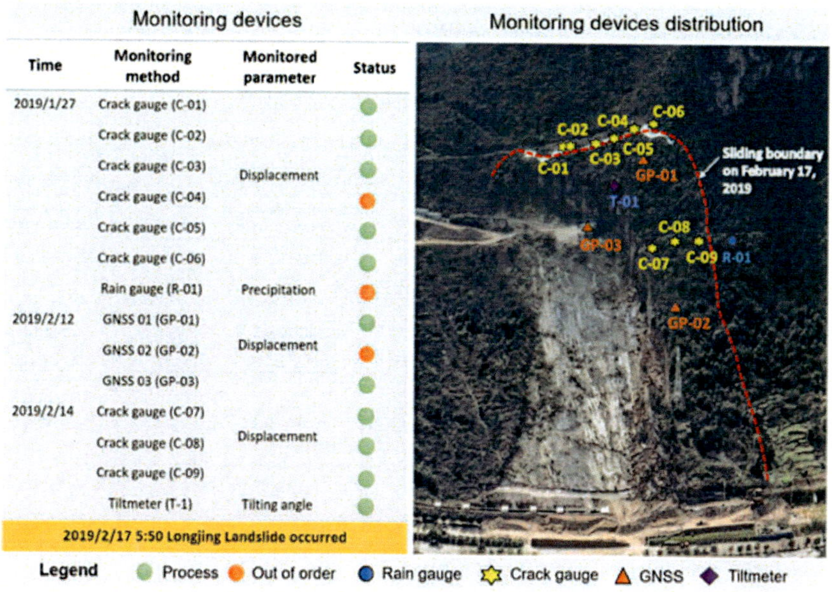

Fig. 3.8 Deployment of monitoring devices on the rocky landslide

alarm information, and the landslide occurred at 5:53:17 am. The LEWS provided a successful early warning inform to the local authorities 53 min in advance.

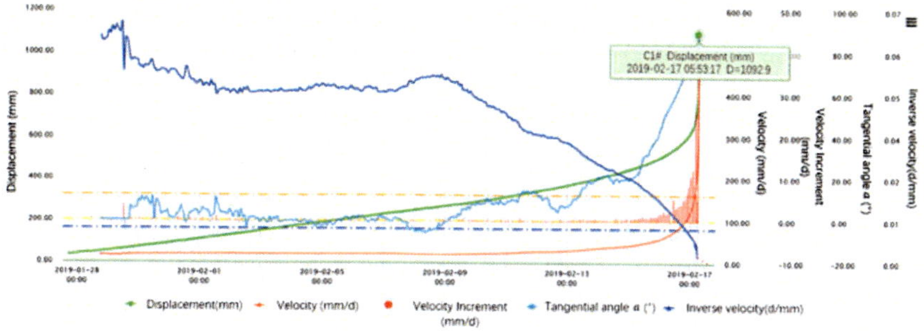

Fig. 3.9 Analysis result of the proposed LEWS based on the monitoring displacement of crack gauge (C-01)

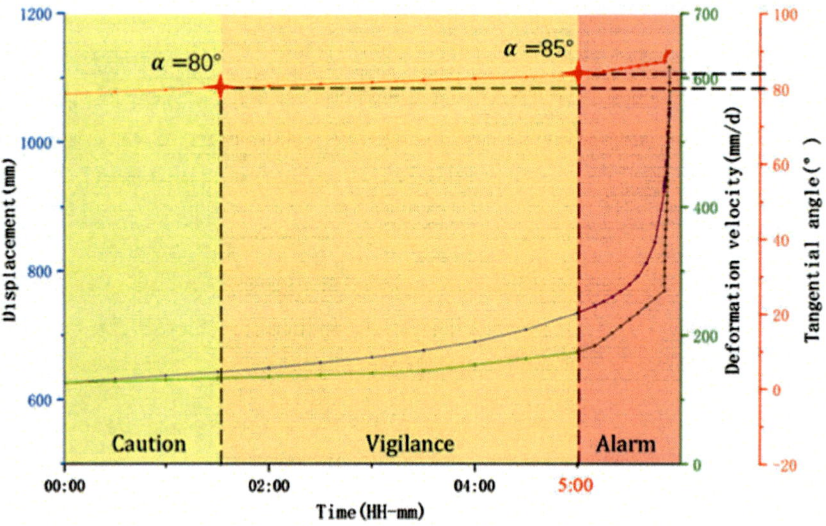

Fig. 3.10 Detailed early warning information during acceleration phase of Longjing rockslide

3.5 Conclusions

The prediction and mitigation of landslide are very challenging worldwide due to its complex nature. An intelligent monitoring and early warning system is a powerful tool for landslide risk reduction. In this chapter, we presented a successful case of early warning for a large disastrous rockslide in Southwestern China. Our developed LEWS helped to predict the large rockslide, eventually achieving zero casualties or injuries and almost no property losses. The main findings can be summarized as following:

(1) A real-time high-quality monitoring technique and deployment strategies based on the understanding of geological settings of landslide are the first step to final successful early warning.
(2) A comprehensive early warning model with multiple alert criteria threshold is the key part of the landslide early warning system. And the developed WebGIS platform embedded with the proposed model can provide a real-time observation and analysis of landslide state.
(3) Actually, this LEWS has successfully warned more than 10 landslides, which includes 8 loess landslides in Gansu province since its implementation in 2017. Those successful cases demonstrated that the LEWS is a powerful and robust tool for landslide risk mitigation and can be regard as a good reference approach.

References

Azzam R, Fernandez-Steeger TM, Arnhardt C, Klapperich H, Shou KJ (2011) Monitoring of landslides and infrastructures with wireless sensor networks in an earthquake environment. In: Proceedings of 5th international conference on earthquake geotechnical engineering

Carlà T, Intrieri E, Di Traglia F, Nolesini T, Gigli G, Casagli N (2017) Guidelines on the use of inverse velocity method as a tool for setting alarm thresholds and forecasting landslides and structure collapses. Landslides 14(2):517–534

Fan X, Xu Q, Liu J, Subramanian SS, He C, Zhu X, Zhou L (2019) Successful early warning and emergency response of a disastrous rockslide in Guizhou province, China. Landslides 16(12):2445–2457

Intrieri E, Carlà T, Gigli G (2019) Forecasting the time of failure of landslides at slope-scale: a literature review. Earth Sci Rev 193:333–349

Ju NP, Huang J, Huang RQ, He CY, Li YR (2015) A Real-time monitoring and early warning system for landslides in Southwest China. J Mt Sci 12(5):1219–1228

Ju N, Huang J, He C, Van Asch TWJ, Huang R, Fan X, Xu Q, Xiao Y, Wang J (2020) Landslide early warning, case studies from Southwest China. Eng Geol 279

Madhusudhan YV, Renuka HC, Bhavya SS, Manjunatha MN (2018) Landslide and rockslide detection system with landslide early warning system for railways. Int J Eng Res Technol 6(13):1–9

Pecoraro G, Calvello M, Piciullo L (2018) Monitoring strategies for local landslide early warning systems. Landslides 16(2):213–231

Petley D (2012) Global patterns of loss of life from landslides. Geology 40(10):927–930

Saito M (1969) Research on forecasting the time of occurrence of slope failure. Soil Mech Found Eng 17(2):29–38

Segalini A, Valletta A, Carri A (2018) Landslide time-of-failure forecast and alert threshold assessment: a generalized criterion. Eng Geol 245:72–80

Tavenas F, Leroueil S (1981) Creep and failure of slopes in clays. Can Geotech J 18(1):106–120

Xu Q, Zeng Y, Qian J, Wang C (2009) Study on an improved tangential angle and the corresponding landslide pre-warning criteria. Geol Bull China 28:501–505

Xu Q, Yuan Y, Hack R (2011) Some new pre-warning critera for creep slope failure. Sci China Technol 54:210–220

Xu Q, Peng D, Zhang S, Zhu X, He C, Qi X, Zhao K, Xiu D, Ju N (2020) Successful implementations of a real-time and intelligent early warning system for loess landslides on the Heifangtai terrace, China. Eng Geol 278

Zhou XP, Liu LJ, Xu C (2020) A modified inverse-velocity method for predicting the failure time of landslides. Eng Geol 268

Zhu X, Xu Q, Qi X, Liu H (2017) A self-adaptive data acquisition technique and its application in landslide monitoring. Springer International Publishing

Chapter 4
Prediction of Slope Stability Using Ensemble Learning Techniques

4.1 Introduction

Landslides are one of the most catastrophic natural hazards occurring in mountainous areas, which have attracted increasing concern in geotechnical and geological practice because they may induce considerably detrimental social and economic impacts (e.g., Huang et al. 2018, 2020a, b; Tang et al. 2019; Wang et al. 2019a, b). China is one of the countries with the most serious landslides in the world. For example, there have been more than 5000 landslides or potential landslides distributed in the Three Gorges Reservoir Area (TGRA) since the first impoundment in 2003 (Gu et al. 2017). Thus, it is of great significance to evaluate the slope stability for designing remedial and mitigation measures. Generally, for a specific slope case, its stability can be rationally and explicitly quantified through performing slope stability analysis with the aid of existing commercial geotechnical software. However, when facing such a larger number of slopes or potential landslides, it may unrealistic to conduct slope stability analysis for all of them in engineering practice. In such as case, slope stability prediction has gained popularity in geotechnical and geological engineering.

In the past few decades, many researchers have contributed to slope stability prediction (e.g., Qi and Tang 2018; Zhou et al. 2019; Pham et al. 2021; Kardani et al. 2021). For example, Qi and Tang (2018) compared the predictive performance of six integrated artificial intelligence approaches in slope stability prediction based on the 168 slope cases collected from the literature. Zhou et al. (2019) introduced the gradient boosting machine method to slope stability prediction using a database that contains 221 different actual slope cases with circular mode failure. Kardani et al. (2021) developed a hybrid stacking ensemble approach for improving the slope stability prediction based on synthetic data and field data. In their studies, the slope height, slope angle, pore water pressure coefficient, unit weight, cohesion, and internal friction angle were used as the main inputs, and the stability status of slopes was regarded as the output. It is well recognized that slope stability is influenced by many factors, such as mechanical parameters, geometric variables, topographic features, and geological conditions. The previous studies mainly focused on the

Z. Wengang et al., *Application of Machine Learning in Slope Stability Assessment*,
https://doi.org/10.1007/978-981-99-2756-2_4

mechanical parameters and geometric variables in the slope stability prediction; in contrast, the topographic features and geological conditions are rarely considered. With the rapid advancement of remote sensing technology and geologic exploration technology, the topographic features and geological conditions for a specified region of interest can be acquired rationally (Huang et al. 2020a, b; Ji et al. 2020). Based on the contribution of previous research, this study tries to reveal the relative importance of topographic features and geological conditions in slope stability prediction using ensemble learning techniques.

Ensemble learning techniques make full use of multiple predictors to form a superior one for improving the performance of machine learning, which has attracted increasing attention in geotechnical engineering applications (e.g., Wang et al. 2020a, b, c; Liu et al. 2021; Zhang et al. 2021a, b; Zhou et al. 2021). Generally, ensemble learning techniques can be broadly categorized into two groups according to their structures, including bagging (parallel) and boosting (sequential) (Zhang et al. 2021a). The well-known RF belongs to the bagging method, which can be it can be regarded as a process of consensus decision tree making and each decision tree decides independently on a specific issue (Breiman 2001). In contrast, extreme gradient boosting (XGBoost) is developed under the boosting framework, where each decision tree learns from its predecessors and updates the residual errors (Chen and Guestrin 2016). The XGBoost has been widely used in the famous Kaggle machine learning competitions due to its advantages of high efficiency and sufficient flexibility.

This study aims to develop an ensemble learning-based approach to predict the stability status of slopes using the RF and XGBoost. As an illustration, the proposed approach is applied to Yunyang County, which records the stability status of 786 historical landslide cases. The remainder of this chapter starts with the introduction of the study area, followed by a brief description of the RF and XGBoost. Finally, the predictive performance of RF, XGBoost, SVM, and logistic regression (LR) in the slope stability prediction of Yunyang County is systematically explored, and the feature importance of the 12 influencing factors is ranked.

4.2 Study Area

The Yunyang County covers an extent of approximately 3649 km^2, which is located in the Chongqing of China. The shoreline of the TGRA in the study area is approximately 707.8 km, accounting for about 11% of the overall length (i.e., 6300 km). To facilitate the understanding of the study area, the following subsections will provide a brief description of the Yunyang County, including topographic conditions, geological conditions, and the features of landslide cases.

4.2.1 Topographic Conditions

Elevation accurately reflects the altitude of a specific geographical location, the elevation raster visualization has been made to obtain the elevation map of Yunyang County using ArcGIS software. According to the investigation report of geological disasters in Yunyang County, this area contains a large number of landslides, and the number of landslides is more in low altitude regions while is less in high altitude regions. By analyzing the front edge elevation and back edge elevation statistics of landslides, the frequency distribution of the front edge elevation and back edge elevation can be obtained, as shown in Fig. 4.1a, b. It can be seen that the height of the deformed slope is mostly between 200 and 700 m, and the slopes with an elevation of about 400 m are the most.

The height partly represents the scale of a slope, which is strongly related to the inner stress of the slope. It plays a controlling role in slope stability. Generally speaking, higher slopes are accompanied by stronger disturbance sensitivity. The

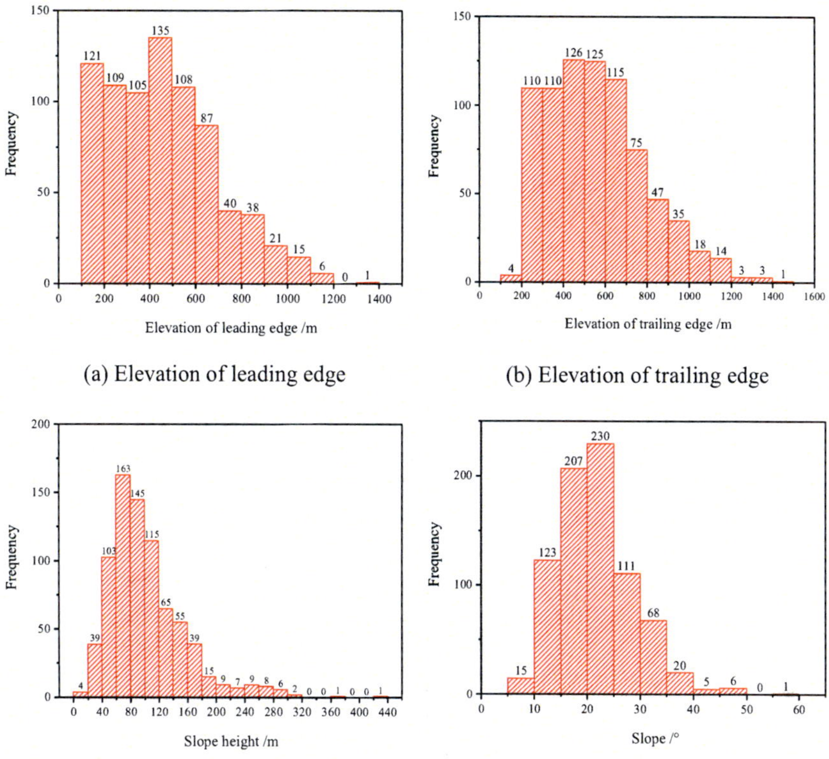

(a) Elevation of leading edge (b) Elevation of trailing edge

(c) Height (d) Slope

Fig. 4.1 Frequency histogram of topographic conditions

combined effect of the height of slope and slope gradient plays a crucial role in slope stability. The height of slopes is obtained by subtracting back edge elevation from front edge elevation. According to the frequency histogram of the height shown in Fig. 4.1c, most of the slopes are between 40 and 120 m high, while the height of slopes higher than 180 m accounts for a smaller proportion.

The slope directly affects the amount of geomaterial deposited on a slope and further affects the stability of the slope. The slope map of the study area can be obtained by ARCGIS software using grid analysis to transform the coordinate system and extract the slope value in the elevation figure. Besides, Fig. 4.1d shows the frequency histogram of slope distribution. It can be observed that the gradient of the slopes is mainly between 10° and 30°, slopes with a gradient more than 30° take up a few parts and only one slope with gradient more than 50°.

4.2.2 Geological Conditions

Due to differences in tectonic movement, different rock masses behave a great discrepancy in cohesion and shear strength, which therefore leads to various evolution trends of slope mass subject to the identical external condition. For example, hard rock mass placed in the same external condition behaves a small deformation or even has no any damage. The statistics show that rock masses in the study area are mainly mudstone, argillaceous limestone, sandstone, sandy mudstone, and shale. According to the qualitative classification table of rock hardness level, slope rock masses in Yunyang County are divided into very soft rock, soft rock, moderately hard rock, and moderately soft rock. Figure 4.2a plots the frequency histogram of lithological property, where the four rock mass types are separately coded as 0 (very soft rock), 1 (soft rock), 2 (moderately hard rock), and 3 (moderately soft rock). The classification statistics of rock weakness degree shows that moderately soft rock poses the highest proportion while very soft rock is the lowest one.

The steeper rock inclination is likely to lead to rock mass suffering weathering, denudation, and deformation failure, which therefore results in changing the slope morphology. Figure 4.5b shows the frequency histogram of rock inclination. It can be observed that the rock inclination is mainly distributed in 5–25°and the number of slope decreases with the increase in rock inclination when it is greater than 10°.

The dip direction is one of three main elements of rock layer attitude, indicating the spatial orientation of a rock layer. The attitudes of rock layers in different regions are varied because of corresponding diverse sedimentary environments and tectonic movements. The classification can be made on the slope structure by considering the intersection angle between the dip angle and the slope angle. There are many types of slope structures because of various possible dip directions of rock layers, as shown in Fig. 4.2c. The statistics show that slope body with north and south dip direction is most common in the study area, which is mainly related to the geographical location of Yunyang County.

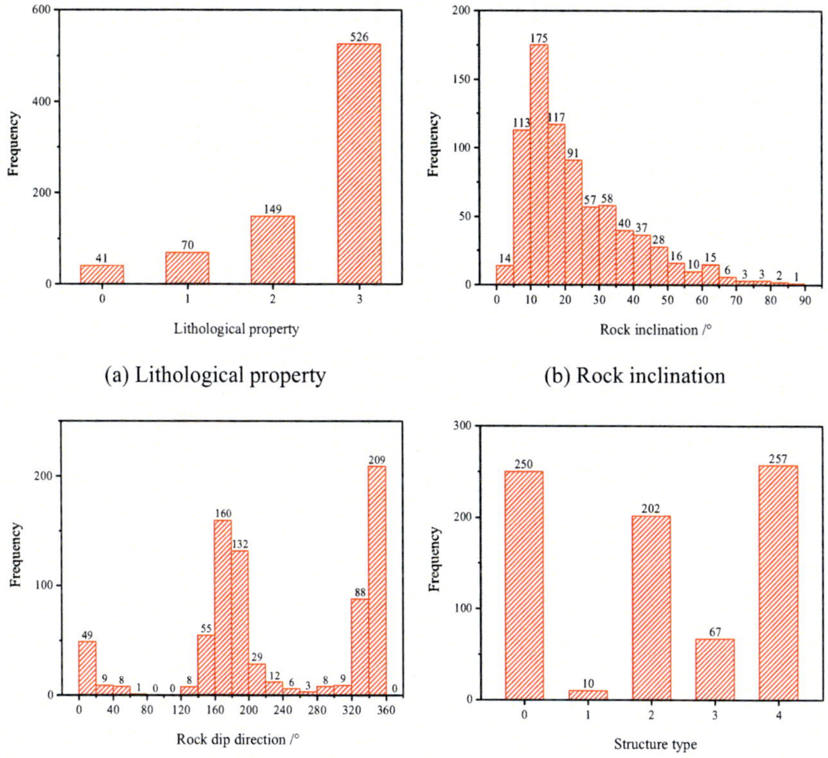

(a) Lithological property (b) Rock inclination

(c) Rock dip direction (d) Structure type

Fig. 4.2 Frequency histogram of geological conditions

The structure reflects the relative position of the rock layer in the slope. According to the combination relationship between the dip angle and the slope angle, the slope can be categorized as dip slope, anti-dip slope, oblique-dip slope, cross-dip slope, and horizontally layered slope. Considering the slope structure type is a categorical variable, the five slope structure types are separately numbered as 0 (anti-dip slope), 1 (horizontal layered slope), 2 (oblique-dip slope), 3 (cross-dip slope), and 4 (dip slope), as plotted in Fig. 4.2d. It is shown that the dip slope, the anti-dip slope, and the oblique-dip slope are the most common types, and the cross-dip slope is the second-largest proportion while the horizontally layered slope is the least common type. Interested readers are referred to Zhang et al. (2022) for more details about the study area.

4.2.3 The Features of Landslide Cases

The plane morphology and profile shape reflect the deformation area and overall shape of the landslide. The statistics of different plane morphology and profile shapes in the study area are carried out, as plotted in Fig. 4.3a, b. Since this parameter is a categorical variable, the number was used to encode each type, where six types of plane morphology are separately coded as 0 (irregular shape), 1 (semicircular shape), 2 (laterally long shape), 3 (rectangle shape), 4 (dustpan shape), and 5 (tongue shape). The tongue shape and irregular shape account for the majority, followed by rectangular, semicircular, laterally long, and dustpan shapes. For the five types of profile shape, the numbers 0, 1, 2, 3, and 4 represent the convex shape, concave shape, composite shape, flat shape, and ladder shape, respectively. As shown in Fig. 4.3b, concave shape, convex shape, and flat shape are the majority, and other forms are relatively small.

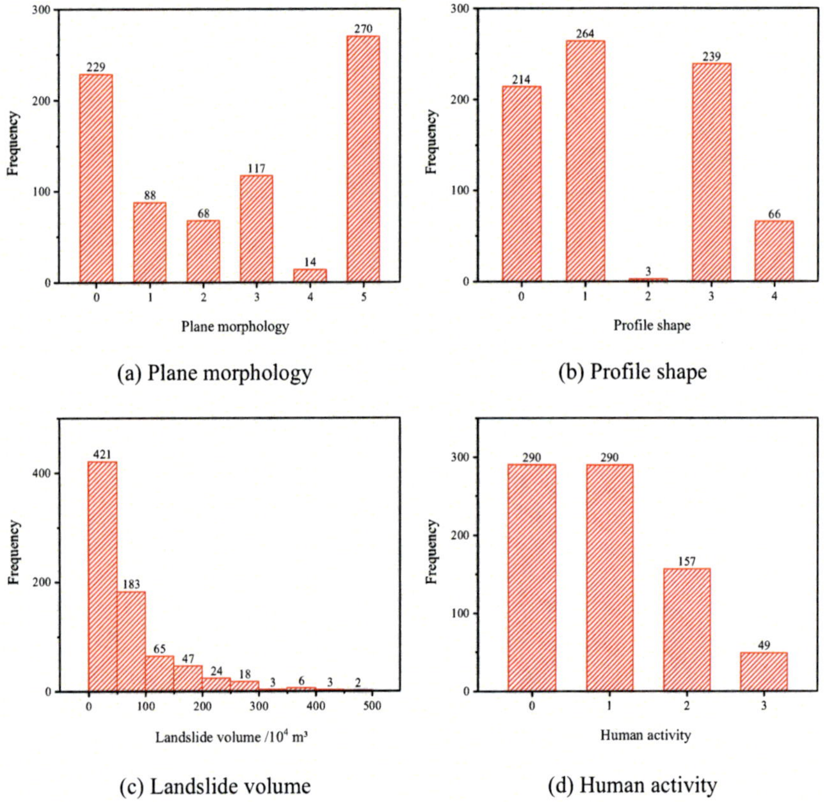

Fig. 4.3 Frequency histogram of slope features

Through the statistics of the landslide volume in the study area, it can be seen from Fig. 4.3c that the slope volume is concentrated in the range of 0–500,000 m^3, accounting for more than half of the total number. The number of landslides gradually decreases with the increase of volume. The engineering construction activities carried out by human beings may exert influence on slope stability. In the study area, the main human activities involve underground excavation, post-slope loading, vegetation destruction, slope cutting, and blasting vibration, which are separately numbered as 0, 1, 2, and 3 in Fig. 4.3d. It is observed that underground excavation and post-slope loading are the major influencing factors.

4.3 Methodology

4.3.1 Extreme Gradient Boosting

The XGBoost uses a gradient boosting framework and is also a decision-tree-based ensemble method (Chen and Guestrin 2016). The core principle of this method is that it builds classification or regression trees one by one, and then the residuals of the previous tree are used to train the subsequent model. It integrates the values calculated from the previously trained tree to achieve a better outcome in the training process. To avoid over-fitting, the pruning procedure is required to be performed, which reduces the size of a decision tree by removing decision nodes that contribute little to target values. The prediction is calculated as follows:

$$\hat{y}_i^{(t)} = \sum_{k=1}^{t} f_k(x_i) = \hat{y}_i^{(t-1)} + f_t(x_i) \tag{4.1}$$

where $\hat{y}_i^{(t)}$ denotes the final tree model; $\hat{y}_i^{(t-1)}$ is the previous tree model; $f_t(x_i)$ is the newly generated tree model and t is the total number of base tree models.

It is important to select appropriate values of depth and number of trees for achieving optimal model performance. Accordingly, the objective function $\text{Obj}^{(t)}$ can be given by:

$$\text{Obj}^{(t)} = \sum_{i=1}^{t} l(y_i, \hat{y}_i^{(t)}) + \sum_{i=1}^{t} \Omega(f_i) \tag{4.2}$$

where y_i is the actual value; $\hat{y}_i^{(t)}$ is the predicted value; $l(y_i, \hat{y}_i^{(t)})$ is the loss function; and $\Omega(f_i)$ is the regularization term.

Following Chen and Guestrin (2016), the objective function can be transformed into the form of the following equation:

$$\text{Obj}^{(t)} = \sum_{i=1}^{t}\left[g_i f_t(x_i) + \frac{1}{2}h_i f_t^2(x_i)\right] + \Omega(f_t) \tag{4.3}$$

where $g_i = \partial y_i^{(t-1)} l(y_i, y_i^{(t-1)})$ and $h_i = \partial^2 y_i^{(t-1)} l(y_i, y_i^{(t-1)})$ are the first- and second-order partial derivatives of the loss function. The regularization term $\Omega(f_t)$ is used to reduce model complexity, avoid over-fitting, and further enhance the generalization ability. It is evaluated by:

$$\Omega(f_t) = \gamma T + \frac{1}{2}\lambda\|w\|^2 \tag{4.4}$$

where T is the number of leaves; w is the corresponding weights of the leaf; λ and γ are coefficients, and the default values are 1 and 0, respectively.

4.3.2 Random Forest

The RF is a tree-based ensemble learning algorithm with the base evaluator of several decision trees (Ho 1995; Breiman 2001). The main idea of RF is to combine the predicted results of many decision trees to provide an ensembled result. In other words, the final prediction can be obtained by averaging the results from all decision trees, which may be more reliable and convincing in many applications compared with a single decision tree. In this study, the base evaluator of several regression trees (RTs) can be further specified. For every branch of an RT, the mean of the samples from the leaf nodes will be calculated. The RTs will continue growing until the mean square error between each sample reaches the minimum or no more features are available. To obtain an ideal RF model, two key parameters should be optimized, namely the number of regression trees (n_estimators) and the maximum depth of the regression tree (max_depth). The RF can be applied to tackle classification and regression problems, which has been widely used in geotechnical engineering with satisfactory performance (Zhang et al. 2021a, c).

4.3.3 Data Preprocessing and Performance Measures

Data preprocessing is an essential task for building a machine learning model, which is because the original database may have missing values, duplicate values, and outliers. In this study, the main steps for data preprocessing are summarized as follows:

Firstly, determine the factors influencing the stability of slopes, in which the continuous and enumerated variables should be distinguished, and at the same time, the text-descriptive variables need to be coded. Among the influencing factors selected in this study, there are seven continuous numerical variables, including the leading elevation, the trailing elevation, the height, the slope, the rock inclination, the rock dip direction, and the landslide volume. In addition, there are five enumerated variables, including lithological property, structure type, plane morphology, profile shape, and human activity. For the continuous numerical variables, they are not further coded as the established model can distinguish their magnitude. For the enumerated variables, since the machine learning model is not distinguishable for the text-descriptive features, each enumerated variable will be coded using numerical value, that is, each enumerated variable is represented by a number.

Secondly, filter the original data samples. In this study, the missing values in the database mainly exist in the feature of profile shape. Considering that the strategy of filling missing values does not necessarily conform to reality and has an impact on the true statistical characteristics of the sample, this chapter directly deletes the data samples with missing values, and finally, there are 786 data samples. It is found that there are no repeated samples in this new database.

For the normalization issue, if the normalization procedure is performed for both continuous numerical variables and enumerated numerical variables, the enumerated numerical variables may deviate from the original enumerated variables. At the same time, as the text-descriptive data has been coded, the text-descriptive variable should not be normalized. When the number of features is relatively large, the feature selection generally needs to be taken into consideration. In this study, through analyzing the influencing factors about the stability of slopes, a total of 12 influencing factors are selected for establishing machine learning models and further assessing the slope stability. To facilitate the coding and analysis, the 12 influencing factors are numbered as listed in Table 4.1.

The database constructed by 786 landslide samples in Yunyang County is randomly divided into two datasets in the ratio of 8:2, and accordingly, 628 data samples are regarded as the training dataset and 158 data samples as the testing

Table 4.1 Number of influencing factors

Impact factor	Number
Elevation of leading edge	F1
Elevation of trailing edge	F2
Height	F3
Slope	F4
Lithological property	F5
Rock inclination	F6
Rock dip direction	F7
Structure type	F8
Plane morphology	F9
Profile shape	F10
Landslide volume	F11
Human activity	F12

Table 4.2 Landslide dataset partition

State	Dataset		
	Training	Testing	Sum
Stable	34	5	39
Basically stable	543	140	683
Less stable	51	13	64
Sum	628	158	786

dataset, as summarized in Table 4.2. The training and testing datasets are successfully randomly divided, and in other words, these two datasets meet the requirement of uniform distribution in the study area.

In the evaluation of model accuracy, the correct rate and error rate are generally applied to assess the model performance. However, these two indicators are not enough to comprehensively describe the model accuracy, and in light of this, the confusion matrix is necessary to be adopted to judge the correct rate and error rate of models for each situation. The confusion matrix is a table used for evaluating the good or poor performance of a classification model, where the columns of the matrix contain the actual classification situations, and the rows of the matrix contain the corresponding predicted classification situations. Each element in the confusion matrix represents the number of the predicted states corresponding to the actual states. Based on the confusion matrix, the indexes of recall rate, precision, and accuracy can be conveniently calculated for evaluating the predictive performance of the established machine learning models.

In this study, a comparative study is conducted to compare the performance of RF, XGBoost, SVM, and LR in the slope stability prediction of Yunyang County. During the establishment of machine learning models, 12 influencing factors are regarded as the input, and the target variable/output is the slope stability state which can be categorized into three groups, namely stable state, basically stable state, and less stable state. Based on the constructed models, the predictive performance of the four machine learning methods is systematically investigated and compared, as discussed in the next section.

4.4 Results and Discussion

The Gaussian radial basis kernel function is adopted in the SVM model construction, and the weight parameter class_weight = 1: 10 is selected for addressing the sample imbalance problem. Figure 4.4 summarized the confusion matrix calculated from the SVM model. Results show that the recall rate of the basically stable state in the training dataset and testing dataset is the same (i.e., 1.000), and the overall accuracy of them is 0.865 and 0.886, respectively. However, for the stable state and less stable state cases, both of them are predicted to be the stable state when using the SVM model. Accordingly, the recall rates of these two cases are 0. This means that although the SVM model can identify the majority class correctly, there are shortcomings in capturing the minority class, which may be attributed to the deficiencies in tackling the imbalanced data problem.

For the RF model, N_estimators and Max_depth are the two most important hyper-parameters. In this study, the grid search method is used to determine the optimal values of these two hyper-parameters, namely N_estimators = 16 and Max_depth = 15. Figure 4.5 lists the confusion matrix evaluated from the RF model. The overall accuracy of the model reaches 0.990 in the training dataset, and the recall rates for the stable state, basically stable state, and less stable state are 0.853, 1.000, and 0.980, respectively. In the testing dataset, the overall accuracy of the RF model is 0.911. In terms of the recall rate, the recall rate of the basically stable state is 1.000. For the stable state cases, all of them are incorrectly judged to be basically stable state. For the less stable state cases, 9 of them are mistakenly identified to be basically stable state. It can be observed that the prediction results of the RF model on the training dataset are relatively satisfactory. In contrast, the RF model simply performs well in judging the basically stable state. Nonetheless, the prediction performance of the RF model is improved compared with the SVM model.

Figure 4.6 tabulates the confusion matrix calculated from the XGBoost model. It is shown that the overall accuracy and recall rate of the three cases of stable state, basically stable state, and less stable state are 1.000 in the training dataset. In the testing dataset, the overall accuracy of the XGBoost model is 0.905, and the recall rate of the basically stable state is 1.000. All the stable state cases are mistakenly judged to be basically stable state, and the proportion of correct judgment for the less stable state case is 0.231. In general, both the two ensemble learning models (i.e., XGBoost and RF) perform better than the SVM model.

Moreover, Fig. 4.7 summarizes the confusion matrix obtained from the LR model. The overall accuracy of the LR model is 0.866 in the training dataset, and the recall

Predicted class (columns) / True class (rows)

	Stable	Basically stable	Less stable	Recall
Stable	0	34	0	0.000
Basically stable	0	543	0	1.000
Less stable	0	51	0	0.000
Precision	0.000	0.865	0.000	0.865

(a) Training dataset

	Stable	Basically stable	Less stable	Recall
Stable	0	5	0	0.000
Basically stable	0	140	0	1.000
Less stable	0	13	0	0.000
Precision	0.000	0.886	0.000	0.886

(b) Testing dataset

Fig. 4.4 Confusion matrix of SVM algorithm prediction value

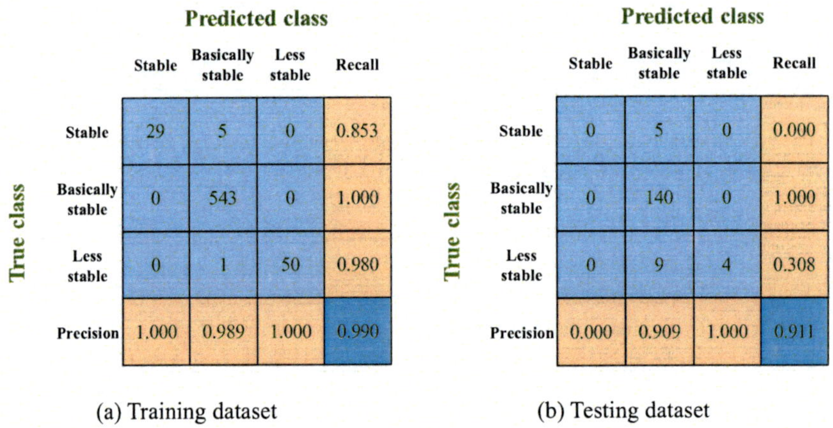

Fig. 4.5 Confusion matrix of RF algorithm prediction value

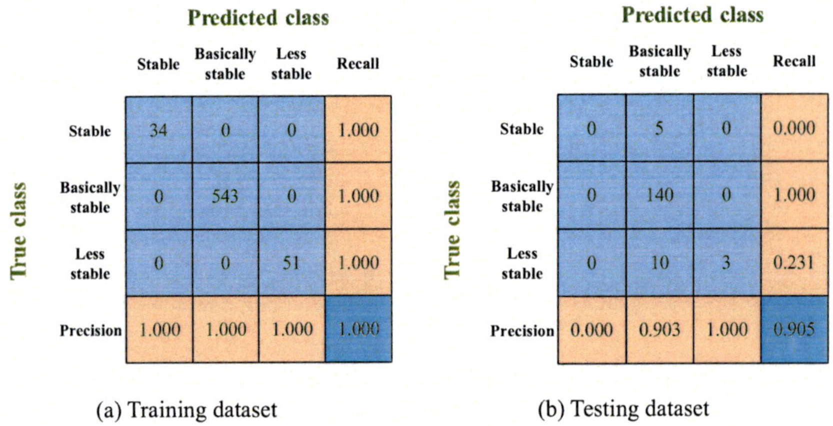

Fig. 4.6 Confusion matrix of XGBoost algorithm prediction value

rate for the basically stable state cases is 1.000. However, all the stable state cases are mistakenly judged to be basically stable state, and only one less stable state case is correctly identified. In the testing dataset, the overall accuracy of the LR model is 0.886, and the recall rate of the basically stable state is 1.000. In contrast, both the recall rates of the stable state and less stable state are 0. Although the LR model performs similarly with the SVM model in the slope stability prediction, both of them lag behind the two ensemble learning models (i.e., XGBoost and RF).

Table 4.3 summarizes the predictive performance of the four machine learning models. In general, the XGBoost model and RF model perform better than the SVM model and LR model in the slope stability prediction of Yunyang County. The main advantages of ensemble learning models over the SVM model and LR model are

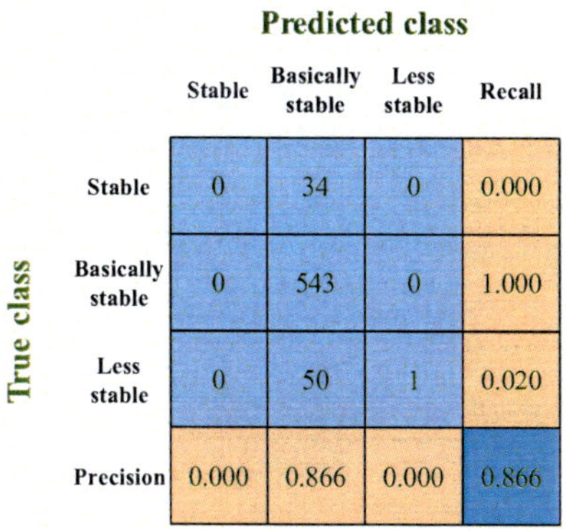

(a) Training dataset

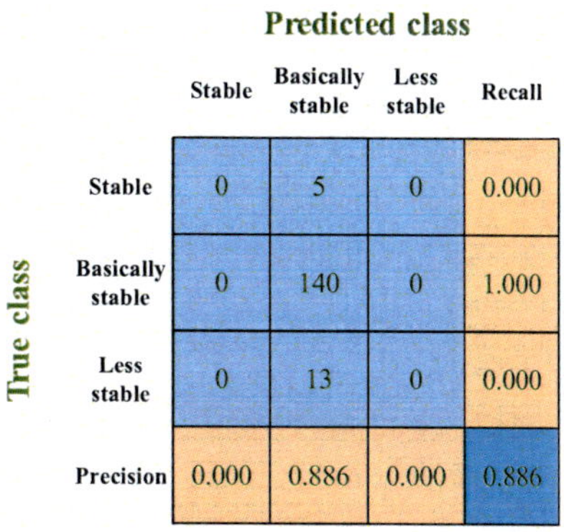

(b) Testing dataset

Fig. 4.7 Confusion matrix of LR algorithm prediction value

Table 4.3 Predictive performance of the four models

Model	Accuracy	
	Training dataset	Testing dataset
SVM	0.865	0.886
RF	0.990	0.911
XGBoost	1.000	0.905
LR	0.866	0.886

their capacity to make full use of multiple predictors to form a better one with high accuracy.

The feature importance is an essential reference for feature selection and model interpretability in machine learning. Figure 4.8 ranks the feature importance of the 12 influencing factors for the XGBoost model. It can be seen that the profile shape is the most important feature variable, accounting for more than 0.1. This implies that the profile shape plays a vital role in the slope stability prediction of Yunyang County.

4.5 Summary and Conclusions

This chapter introduced two promising ensemble learning called RF and XGBoost to predict the stability status of slopes. For illustration, the proposed approach was applied to Yunyang County, which records the stability status of 786 historical landslide cases. A comparative study was conducted to compare the predictive performance of RF, XGBoost, SVM, and LR. Moreover, the feature importance of the 12 influencing variables (i.e., elevation of leading edge and trailing edge, height,

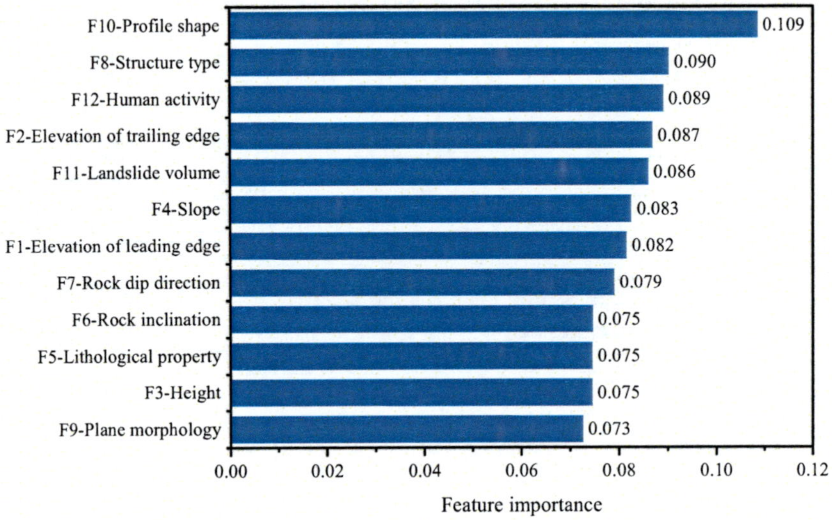

Fig. 4.8 Feature importance of impact factors

slope, lithological property, rock inclination, rock dip direction, structure type, plane morphology, profile shape, landslide volume, and human activity) was also explored using the results obtained from the XGBoost model.

Results showed that the accuracy of the XGBoost, RF, LR, and SVM on the training dataset was 1.000, 0.990, 0.866, 0.865, respectively, and the accuracy of those on the testing dataset was 0.905, 0.911, 0.886, 0.886, respectively. The obtained results revealed the superiority of ensemble learning models (i.e., the XGBoost and RF) over the conventional SVM model and LR model in the slope stability prediction of Yunyang County. The main advantages of ensemble learning models over the SVM model and LR model are their capacity to make full use of several predictors to form a better one with high accuracy. It provides a possibility of integrating ensemble learning techniques into slope stability prediction for rationally capturing the stability status of slopes in geotechnical and geological applications. Among the 12 influential factors, the profile shape was found to be the most influential variable in the slope stability prediction. Last but not least, it is worthy pointing out that Yunyang County was used in this chapter for illustration, and the proposed ensemble learning-based approach can also be extended to other landslide-prone regions provided that the corresponding database needed for the model calibration is available.

References

Breiman L (2001) Random forests. Mach Learn 45(1):5–32

Chen T, Guestrin C (2016) Xgboost: a scalable tree boosting system. In: Proceedings of the 22nd acm sigkdd international conference on knowledge discovery and data mining. ACM, pp 785–794. https://doi.org/10.1145/2939672.2939785

Gu DM, Huang D, Yang WD, Zhu JL, Fu GY (2017) Understanding the triggering mechanism and possible kinematic evolution of a reactivated landslide in the Three Gorges Reservoir. Landslides 14:2073–2087. https://doi.org/10.1007/s10346-017-0845-4

Ho TK (1995) Random decision forests. In: Proceedings of 3rd international conference on document analysis and recognition, vol 1. IEEE, pp 278–282

Huang D, Gu DM, Song YX, Cen DF, Zeng B (2018) Towards a complete understanding of the triggering mechanism of a large reactivated landslide in the Three Gorges Reservoir. Eng Geol 238:36–51. https://doi.org/10.1016/j.enggeo.2018.03.008

Huang D, Luo S, Zhong Z (2020a) Analysis and modeling of the combined effects of hydrological factors on a reservoir bank slope in the Three Gorges Reservoir area, China. Eng Geol 279:105858. https://doi.org/10.1016/j.enggeo.2020.105858

Huang F, Cao Z, Guo J, Jiang SH, Li S, Guo Z (2020b) Comparisons of heuristic, general statistical and machine learning models for landslide susceptibility prediction and mapping. CATENA 191:104580. https://doi.org/10.1016/j.catena.2020.104580

Ji S, Yu D, Shen C, Li W, Xu Q (2020) Landslide detection from an open satellite imagery and digital elevation model dataset using attention boosted convolutional neural networks. Landslides 17(6):1337–1352. https://doi.org/10.1007/s10346-020-01353-2

Kardani N, Zhou A, Nazem M, Shen SL (2021) Improved prediction of slope stability using a hybrid stacking ensemble method based on finite element analysis and field data. J Rock Mech Geotech Eng 13(1):188–201. https://doi.org/10.1016/j.jrmge.2020.05.011

Liu Z, Gilbert G, Cepeda JM, Lysdahl AOK, Piciullo L, Hefre H, Lacasse S (2021) Modelling of shallow landslides with machine learning algorithms. Geosci Front 12(1):385–393. https://doi.org/10.1016/j.gsf.2020.04.014

Pham K, Kim D, Park S, Choi H (2021) Ensemble learning-based classification models for slope stability analysis. CATENA 196:104886. https://doi.org/10.1016/j.catena.2020.104886

Qi C, Tang X (2018) Slope stability prediction using integrated metaheuristic and machine learning approaches: a comparative study. Comput Ind Eng 118:112–122. https://doi.org/10.1016/j.cie.2018.02.028

Tang H, Wasowski J, Juang CH (2019) Geohazards in the three Gorges Reservoir Area, China—Lessons learned from decades of research. Eng Geol 261:105267. https://doi.org/10.1016/j.enggeo.2019.105267

Wang LQ, Yin YP, Huang BL, Zhang ZH, Wei YJ (2019a) Formation and characteristics of Guang'an Village landslide in Wuxi, Chongqing, China. Landslides 16(1):127–138. https://doi.org/10.1007/s10346-018-1086-x

Wang LQ, Yin YP, Zhang ZH, Huang BL, Wei YJ, Zhao P, Hu MJ (2019b) Stability analysis of the Xinlu Village landslide (Chongqing, China) and the influence of rainfall. Landslides 16(10):1993–2004. https://doi.org/10.1007/s10346-019-01240-5

Wang L, Wu C, Tang L, Zhang W, Lacasse S, Liu H, Gao L (2020a) Efficient reliability analysis of earth dam slope stability using extreme gradient boosting method. Acta Geotech 15(11):3135–3150. https://doi.org/10.1007/s11440-020-00962-4

Wang MX, Huang DR, Wang G, Li DQ (2020b) SS-XGBoost: a machine learning framework for predicting newmark sliding displacements of slopes. J Geotech Geoenviron Eng 146(9):04020074. https://doi.org/10.1061/(asce)gt.1943-5606.0002297

Wang HL, Yin ZY, Zhang P, Jin YF (2020c) Straightforward prediction for air-entry value of compacted soils using machine learning algorithms. Eng Geol 279(1):105911. https://doi.org/10.1016/j.enggeo.2020.105911

Zhang WG, Wu CZ, Zhong HY, Li YQ, Wang L (2021a) Prediction of undrained shear strength using extreme gradient boosting and random forest based on Bayesian optimization. Geosci Front 12(1):469–477. https://doi.org/10.1016/j.gsf.2020.03.007

Zhang W, Wu C, Li Y, Wang L, Samui P (2021b) Assessment of pile drivability using random forest regression and multivariate adaptive regression splines. Georisk 15(1):27–40. https://doi.org/10.1080/17499518.2019.1674340

Zhang W, Li H, Li Y, Liu H, Chen Y, Ding X (2021c) Application of deep learning algorithms in geotechnical engineering: a short critical review. Artif Intell Rev. https://doi.org/10.1007/s10462-021-09967-1

Zhang W, Li H, Han L, Chen L, Wang L (2022) Slope stability prediction using ensemble learning techniques: a case study in Yunyang County, Chongqing, China. J Rock Mech Geotechn Eng. https://doi.org/10.1016/j.jrmge.2021.12.011

Zhou J, Li E, Yang S, Wang M, Shi X, Yao S, Mitri HS (2019) Slope stability prediction for circular mode failure using gradient boosting machine approach based on an updated database of case histories. Saf Sci 118:505–518. https://doi.org/10.1016/j.ssci.2019.05.046

Zhou J, Qiu Y, Armaghani DJ, Zhang W, Li C, Zhu S, Tarinejad R (2021) Predicting TBM penetration rate in hard rock condition: a comparative study among six XGB-based metaheuristic techniques. Geosci Front 12(3):101091. https://doi.org/10.1016/j.gsf.2020.09.020

Chapter 5
Landslide Susceptibility Research Combining Qualitative Analysis and Quantitative Evaluation: A Case Study of Yunyang County in Chongqing, China

5.1 Introduction

Landslides are one of the most destructive geological hazards, which not only cause enormous damage to houses and infrastructure, such as bridges and roads, but also lead to loss of life (Huang et al. 2017). According to the World Health Organization, approximately 4.8 million people were affected, and more than 18,000 deaths were caused by landslides between 1998 and 2017. Specifically, as one of the countries with a high incidence of landslides, China suffered severe loss of life (Petley 2012; Lin et al. 2017). The China Statistical Yearbook indicates that during 2000 to 2015, 373,630 landslides occurred in this country, killing 10,996 people, which is approximately 690 landslide-related deaths per year (Sheng et al. 2016). To mitigate the serious social impact caused by landslides, constructive and productive activities should be avoided in areas with high susceptibility to landslides. Therefore, developing an efficient method to distinguish landslide-prone zones is an essential need for both local governments and research institutes (Sun et al. 2021). Landslide susceptibility describes the likelihood that a landslide will occur in a certain area based on local terrain conditions (Brabb 1984). Landslide susceptibility mapping (LSM) is one of the most widely used assessment methods, and it visualizes the spatial distribution of zones with different probabilities of occurrence of landslides in a certain area.

Various methods such as probabilistic analysis, statistical analysis, analytic process, and weighted overlay were widely applied to LSM by researchers in the early stages. With the development of AI and GIS, machine learning-based methods, with the capability of solving complex nonlinear problems, are becoming increasingly popular compared to opinion-driven models and statistical learning, making the accuracy and precision of susceptibility models evolve rapidly (Wu et al. 2022; Wang et al. 2022). Huang et al. (2022) adopted LR, SVM, and RF on LSM for model comparison. He et al. (2021) used RF in the global assessment of earthquake-induced landslide susceptibility. Sun et al. (2020) applied the Bayes algorithm to optimize the hyper-parameters of the RF model for LSM. Smith et al. (2021) compared the

Z. Wengang et al., *Application of Machine Learning in Slope Stability Assessment*,
https://doi.org/10.1007/978-981-99-2756-2_5

effect of landslide inventories assembled by different methods on the performance of RF and LF for LSM. Lim et al. (2022) applied the RF model to estimate the probability of a landslide. Nhu et al. (2020) investigated and compared the logistic model tree, LR, Naive Bayes tree, ANN, and SVM in the shallow landslide susceptibility mapping for Bijar City in Kurdistan City. Zhang et al. (2022a) used the predictive performance of RF, XGBoost, SVM, and LR on landslide susceptibility mapping in Yunyang County. Hu et al. (2020) compared the effect of different non-landslide sampling methods on the performance of SVM and NB for LSM. Zhou et al. (2021) applied GeoDetector and RFE for factor optimization and then used the selected factors as inputs to train an RF model to obtain the LSM of Wuxi County. Sun et al. (2022) proposed a hybrid landslide warning model based on RF susceptibility zoning and precipitation. Zhou et al. (2022) constructed an interpretable model for the susceptibility to rainfall-induced shallow landslides based on Shapley additive explanation (SHAP) and XGBoost.

Among those methods, RF is the most commonly used method in large-scale mapping and classification (Akar and Güngör 2015; Hengl et al. 2015; Feng et al. 2015) due to its characteristics of low computational cost, low data requirement, convenience of hyper-parameters tuning, and robustness in solving complex nonlinear problems (Yin et al. 2022). Previous work usually focused on quantitative analysis, such as the selection and improvement of models and input features, but rarely took into account the qualitative analysis of landslide areas. Actually, as one of the major geological hazards, landslides are highly area dependent, the mechanism of landslide formation and its corresponding triggering factors are undoubtedly different in distinct areas. The frequent fluctuation of reservoir water seriously reduces the stability of the slopes in the reservoir area, making them prone to landslides (Yin et al. 2022). For mountainous areas, however, rainfall is the major triggering factor for the occurrence of landslides (Wang et al. 2021a, b). With increased population, human activities have become the major issue that accelerates landslide formation in areas with high population density. Therefore, manually dividing a relatively large region into different sub-zones according to the qualitative analysis of the landslide formation and geomorphic unit characteristics will theoretically improve mapping accuracy. This part aims to use the 827 historical landslide data points in Yunyang County and the 20 conditioning factors to build 5 RF models, including an RF model (referred to as the parent model below) for the whole region and four RF models for the divided four sub-zones (referred to as sub-model one to sub-model four below). Then, the feature importance and the performances of the parent model and the four sub-models are analyzed and compared to verify the effectiveness of applying experience-based zonation before modeling.

5.2 Study Area

Chongqing City is located in the mountainous area around the eastern Sichuan basin and the slope area of the basin margin. It spans two tectonic units, namely the Yangtze quasi-platform and the Qinling fold system. The landscape of Chongqing City is mainly mountains and hills, which make up 92% of its total area. There are many adverse geological conditions accelerating the formation of landslides, dangerous rock collapse, ground collapse, debris flow, and other geological disasters, including developed surface water networks, strongly cut terrain, complex rock and soil structure, and geological structure, making it one of the cities with the highest geological disaster frequency in the country.

The spatial distribution of geological hazards in Chongqing City shows a certain degree of concentration and can be concluded as a striped distribution and vertical zonal distribution. Moreover, its temporal distribution presents a seasonal cluster pattern. According to the statistics, there are currently 14,926 geological hazard-prone points in Chongqing City, of which 5776 (38.7%) are located in the seven districts and counties of northeast Chongqing City (Wanzhou, Kaizhou, Chengkou, Wuxi, Wushan, Fengjie, Yunyang), 1864 (12.49%) are in the five districts and counties of southeastern Chongqing City (Wulong, Youyang, Qianjiang, Pengshui, Xiushan), and 1320 (8.84%) are in the 11 districts of the main city. Therefore, northeast Chongqing City is the key area with a high probability of potential geological disasters.

As one of the seven districts and counties in the northeast of Chongqing City, Yunyang County (spans 108° 24′ 37″–109° 14′ 47″ E and 30° 34′ 59″–31° 26′ 28″ N) is located in the middle of the Three Gorges Reservoir Project area, being the important hub of the ecological and economic zone along the Yangtze River. According to the announcement of the Chongqing Forest Bureau, while the forest area of Chongqing city reaches 54.5%, that of Yunyang County exceeds 58.5%, making it one of the greenest counties in China. Based on the Seventh National Census of China, there were 929,034 long-term residents (48% of them are urban residents) in this area in the year 2020. Yunyang County is crossed by twelve major folds, namely Changdianfang Syncline (1), Macaoba Anticline (2), Qvmahe Syncline (3), Tiefengshan Anticline (4), Yangliuwan Syncline (5), Dongcun Anticline (6), Xinchang Anticline (7), Huangpoxi Syncline (8), Guling Syncline (9), Fangdoushan Anticline (10), Ganchang Syncline (11), and Longjukan Syncline (12). Under the subtropical monsoon climate, Yunyang County has an average annual rainfall of 1123.7 to 1264.8 mm and an average annual temperature from 10.2 to 18.5 °C.

Mountainous areas are generally susceptible to mass movements due to preparatory and triggering causal factors (Nakileza and Nedala 2020); not only the weathering effects but anthropogenic activities in the region also commonly accelerate the formation of unstable areas on both the earth material and on hill slopes (Nefeslioglu et al. 2011). As a part of Chongqing City, Yunyang County has always been a significant hotspot for landslide occurrences. There are a total of 836 historical landslides recorded in the dataset; 827 data points are left after data cleaning. A total of 28.2% of them are small landslides, 51.8% are medium landslides, and 20% are large landslides. Among them, trust-load-caused landslides accounted for 53.7%, and loosen-caused landslides and multi-caused landslides accounted for 14.5% and 31.8%, respectively. To build sub-models, we manually divided the study area into four different sub-zones based on the information from the exploration of geological hazards in Chongqing City, such as the mechanism of landslide formation and sliding failure and the geomorphic unit characteristics. Among the four sub-zones, sub-zone II contains all the strip-distributing landslides along the mainstream of the Yangtze River, so it can also be called the Yangtze River mainstream zone. From a larger scope, a part of Yunyang County belongs to the low-hills section that crosses Yunyang, Fengjie, and Kaizhou; this area is classified as sub-zone IV. Sub-zone I (south of sub-zone II) is crossed by the main highway called S305, and the main area of sub-zone III (between sub-zone II and IV) is crossed by the S103 and S305. Similarly, the density of the road network is also at a high stage in the other two parts of sub-zone III. The landslides that occurred in these two sub-zones are found to be mainly along the roads. After zonation, 89 landslides are located in sub-zone I, 285 of them are in sub-zone II, sub-zone III contains 44 landslides, and with the largest area, 408 of the historical landslides occurred in sub-zone IV.

As one of the typical landslides in the TGRA, the Jiuxianping landslide (in sub-zone II) is located on the left bank of the Yangtze River. After the Three Gorges Reservoir project, the fluctuation of the Three Gorges Reservoir water level restarted the displacement and deformation of the ancient landslide, making this area more prone to geological hazards. A subsidence of about one meter occurred on a roadway in the middle of the landslide body after heavy rain in 2003 and 2004, causing the roadway to be abandoned. With the impact of continuous heavy rain, landslides occurred in the back accumulation of Jiuxianping on June 19 and 22, 2007, causing the houses of the villagers to collapse, and the mountain body cracked. On June 9, 2009, the back-accumulation of Jiuxianping deformed again under the impact of heavy rain, causing cracks on both the accumulation body and the houses of the villagers. Recently, under the continuous effect of the Three Gorges Reservoir, this area has been in the overall creep deformation stage for years, especially the cliffs near the river, which often suffer from local collapse and damage.

The continuous heavy rains from August 30 to September 1, 2014, made the accumulated rainfall in Jiangkou Town more than 300 mm. The day after that, the Tuantan landslide occurred on the back mountain and on the left side of the Yongfa Coal Mine staff dormitory in Tuantan village, Jiangkou Town, Yunyang County (in sub-zone IV). Although the employees were notified to evacuate from the area subjected to the massive landslide, twelve of them were buried on the spot. Unfortunately, only one of the twelve was saved.

Typically, in sub-zone I, under the impact of rainfall, a landslide occurred on the S202 Highway in the direction from Longjiao to Rucao on July 13, 2021. Similarly, there was a 10,000 cubic-meter landslide triggered by heavy rainfall in the area of Mawang Temple, which trapped two four-wheel cars and a motorcycle, and blocked the highway section for five days. Interested readers are referred to Zhang et al. (2022b) for more details about the study area.

5.3 Method Explanation

5.3.1 Random Forest

As one of the most popular classification methods, RF was first proposed by Breiman (1996) and Cutler and Stevens (2020). During its training process, different data subsets obtained by random sampling are used to train multiple decision trees as independent estimators; each of them is only allowed to fit the data based on the part of the input features, and the final output of the RF model is based on the voting results of the constructed estimators. With the double randomness, those estimators will be trained as distinct ones, and such a special structure makes RF less sensitive to noise and outliers, less likely to overfit and has a lower dependency on feature selection, but more robust and can provide more accurate predictions compared with the other machine learning models.

The RF training process can be concluded as the flowing three simple steps (Fig. 5.1):

Bootstrapping: To build M decision trees (estimators), M subsets will be generated as training sets from the original dataset by sampling with replacement.

Modeling: Each subset obtained by bagging is used to train the corresponding estimator. Similarly, for the purpose of getting estimators as distinct as possible, the available features for each estimator are also randomly chosen from the entire feature list.

Voting: Each estimator will output a result independently, then the final result of the model will be generated based on the voting results.

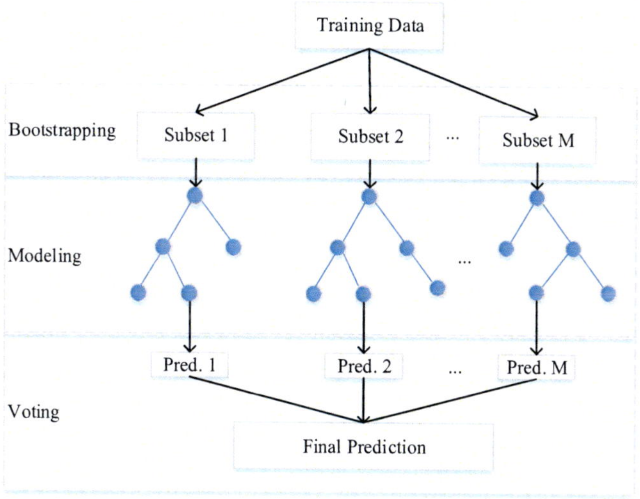

Fig. 5.1 Random forest flowchart

5.3.2 *Grid Search*

As pre-set parameters, hyper-parameters play a crucial role in model performance, and there are rare algorithms that are hyper-parameter-free (Zhang et al. 2020). Therefore, proper hyper-parameter tuning is essential for improving model performance. Grid search, as one of the conventional automatic hyper-parameter tuning methods, is widely used because of its simple operation. Its basic idea is to choose the best hyper-parameter combination by enumerating and iterating over all possible combinations. Although it is computationally expensive because of the exhaustive search process, grid search suits random forest very well, as random forest is a hyper-parameter tuning friendly model; only two hyper-parameters are to be tuned in our case.

5.3.3 *Performance Measure*

For classification problems, accurate-oriented modeling sometimes ends up with "rabid" models that tend to classify all samples as a certain type, especially for unbalanced datasets. Therefore, the confusion matrix is introduced in this case (Table 5.1).

According to Table 5.1, the true positive rate, also called the recall, is defined as TPR = TP/(TP + FN), and the false positive rate is defined as FPR = FP/(FP + TN). After model establishment, its corresponding TPR versus FPR at different cutoff values can be plotted, which is called as receiver operation characteristic (ROC) curve and can be used to represent the predictive ability of the model. The area under the ROC curve (AUC) is commonly used as an index that represents the true classification accuracy of the model.

Table 5.1 Confusion matrix

Predicted values	Actual values	
	Positive	Negative
Positive	True positive (TP)	False positive (FP)
Negative	False negative (FN)	True negative (TN)

5.4 Methodology

5.4.1 Data Collection and Preparation

Data collection and data cleaning are of the top importance in machine learning applications; high-quality data is the foundation of accurate prediction models. Model perfection also plays a vital role; choosing appropriate hyper-parameters will help models to extract useful information from input data more effectively and precisely. In this section, we will introduce the procedure for data preparation and model establishment. The analysis of the performance and the factor importance of different models will be presented in the following sections. The flowchart for this study is shown in Fig. 5.2.

Triggered by the joint effects of natural factors and human activities, the mechanism of the occurrence of landslides is very complex (Sun et al. 2021). Based on the assumption that future landslides will occur under the same conditions as past landslides (Lee and Pradhan 2007), evaluating these factors and analyzing their relationships with historically recorded landslides can contribute to forming the basis for the prediction of future landslides in an area (Raghuvanshi et al. 2015;

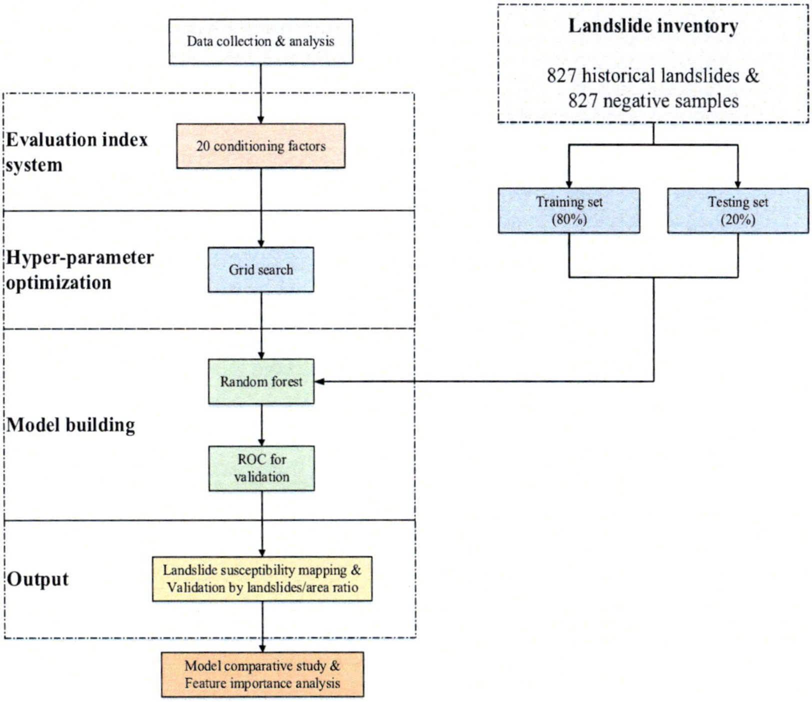

Fig. 5.2 Flowchart of this study

Girma et al. 2015). However, after reviewing the studies related to the evaluation of landslide sensitivity from 1983 to 2016, Reichenbach et al. (2018) found that there was a total of 596 different factors that affected landslide formation, considering all of them will be laborious and time-consuming. According to previous articles, topography, hydrology, geology, land cover, and natural and human-related factors are generally used for landslide susceptibility analysis (Nhu et al. 2020; Zhou et al. 2021). Therefore, in this study, topographical factors (aspect, elevation, plane curvature, profile curvature, relief amplitude, slope), hydrological factors (aridity, distance from rivers, index of moisture (IM), and topographic wetness index (TWI)), geological factors (lithology, distance from anticline axis, distance from syncline axis), factors related to land cover (namely land use and normalized difference vegetation index (NDVI)), human factors (such as human activity intensity of land surface (HAILS), distance from road, and population), and natural factors (average annual temperature, average annual rainfall) are all considered (Fig. 5.3).

As the grid data layers cannot be directly used for model training, the Yunyang County fishnet with a cell size of 25 × 25 m was created for the purpose of data extraction for model training and measuring the distances from specific structures/natural sources, the total number of cells is 5,831,382. The secondary data includes aspect, plane curvature, profile curvature, relief amplitude, slope, distance from rivers, TWI, distance from anticline axis, distance from syncline axis, and distance from road. Among them, TWI, aspect, plane curvature, profile curvature, relief amplitude, and slope were obtained by ArcGIS processing of DEM. The characteristics associated with distances (i.e., distance from rivers, distance from anticline axis, and distance

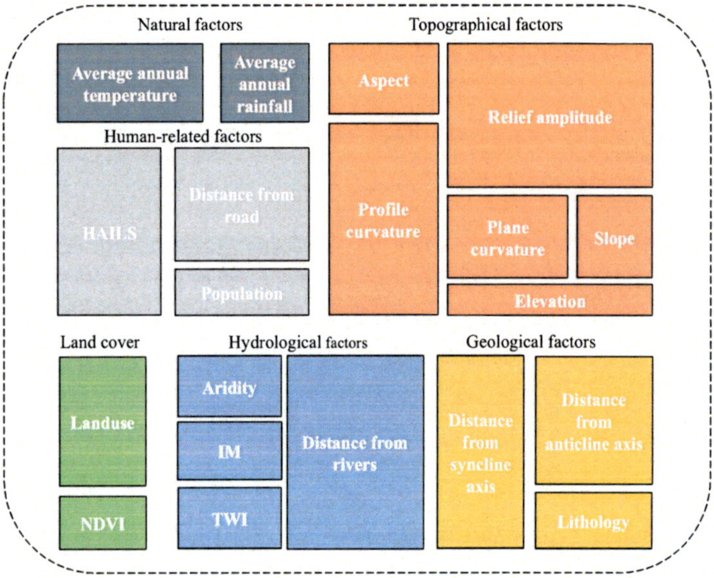

Fig. 5.3 Conditioning factor types

from syncline axis) were obtained by applying the near function of ArcGIS to the corresponding primary data (i.e., the river network of Chongqing City, the road network of Chongqing City, and the geological structure of Yunyang County) and the created fishnet of Yunyang County.

All 20 factors were visualized by ArcGIS with a resolution of 25 m. Then the corresponding classes and rating values of the 20 factors were assigned to each cell of the prepared fishnet. After obtaining the 827 historical landslide cells in the study area as positive samples (1), the same number of non-landslide cells was randomly extracted from the landslide-free areas as negative samples (0) (Yesilnacar and Topal 2005; Mathew et al. 2008). To make sure that the sub-models can be constructed with the same data as the parent model, the number of negative samples selected from each sub-zone was determined by the number of the positive samples in the sub-zone (i.e., 89 for sub-zone I, 286 for sub-zone II, 44 for sub-zone III, and 408 for sub-zone IV). Thus, the five sample datasets were formed. Based on previous research (Nhu et al. 2020; Wang et al. 2021a, b), 80% of each dataset was randomly selected to train the corresponding statistical model, and the remaining 20% was used for validation purposes.

5.4.2 Model Development and Application

Five independent random forest-based models were established and validated based on the sample datasets constructed before. The hyper-parameters of each model were tuned by the grid search method. Being considered as the most straightforward optimization method (LaValle and Branicky 2004), the grid search method needs to iterate over the entire interval of each hyper-parameter. Therefore, the number and the iterating interval of the tuned hyper-parameters have an obvious impact on its search efficiency. The number of estimators determines the stability of a random forest model, and adding more estimators will lower its mean squared prediction error and hence improve model stability (Liu et al. 2017), but will increase computational cost (Lujan-Moreno et al. 2018). The maximum depth of a tree controls the stability of a random forest model in a different way. The stability will decrease as depth increases since increasing the depth will make the model tend to just memorize the training data, but if the forest is too shallow, the model will underfit, resulting in low AUC (Liu et al. 2017). In this case, the number of estimators and the maximum depth of a tree are the two hyper-parameters to be tuned; the results are shown in Table 5.2.

After obtaining well-trained models, we used all 5,831,382 cells prepared in Sect. 4.1as input to generate the landslide susceptibility maps for the study area.

Table 5.2 Hyper-parameters of models

Model	Number of estimators	Maximum depth of a tree
Parent model	175	21
Sub-model one	10	6
Sub-model two	95	12
Sub-model three	10	4
Sub-model four	100	12

5.5 Results

The landslide susceptibility maps are generated by the parent model and sub-models. The entire region is divided and classified into five zones of susceptibility to different levels of landslides (very low, low, moderate, high, and very high) by the method of natural breaks method.

Logically, the landslides/area ratio should increase from a very low landslide-prone zone to a very high landslide-prone zone, which is exactly what our models indicate. According to the results of the parent model, the landslides/area ratio increases from 0.016 to 4.548, from a very low landslide-prone zone to a very high landslide-prone zone, and that also increases from 0.007 to 4.051 from the results of the sub-models. Figure 5.4 displays such a tendency, and it can be seen that the outputs of the sub-models have more obvious gaps between the very low landslide-prone zone and the very high landslide-prone zone, which reflects another merit of the sub-models compared with the parent model.

The validation AUC values of the five models are shown in Fig. 5.5; the AUC value of the parent model is 0.872, while that of sub-model one to sub-model four are 0.949, 0.892, 0.889, and 0.951, respectively. All of the sub-models outperformed the

Fig. 5.4 Landslides/area ratio

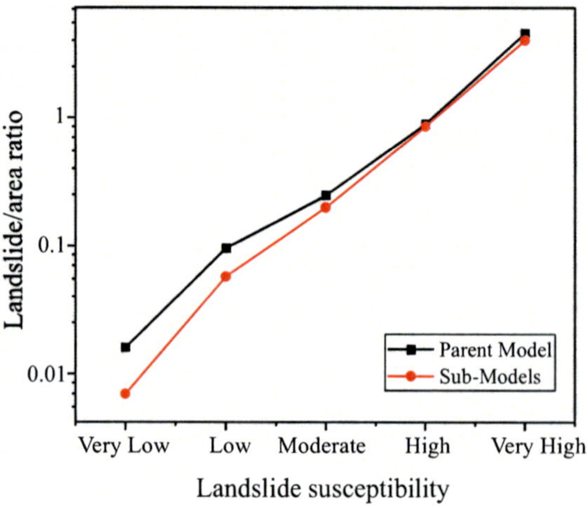

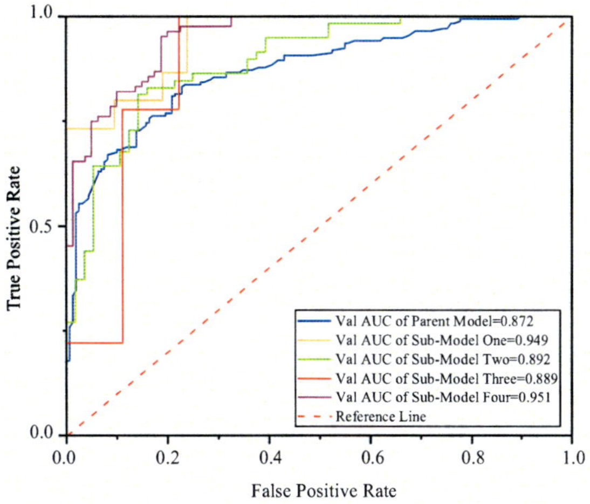

Fig. 5.5 Validation AUC values of different models

parent model, which proved the aforementioned hypothesis. However, the increases are not at the same level; sub-model four achieved the highest improvement (9.1%), while the lowest improvement is 1.9%, which was obtained by sub-model two.

The importance of features generally represents how much a specific feature contributes to the decision-making process of a model. In this case, the most important features can be the key factors in identifying landslide/non-landslide points. As shown in Fig. 5.6, the distance from syncline axis, aridity, elevation, distance from rivers, and average annual temperature is of the highest importance for the parent model. NDVI, average annual rainfall, distance from road, and elevation are the main features that are associated with the formation of landslides in sub-zone I. For sub-model two, elevation is the most important feature, which is followed by average annual rainfall, distance from rivers, distance from anticline axis, and average annual temperature. Distance from road, HAILS, elevation, plane curvature, and average annual temperature are the top features for sub-model three. Last but not least, distance from syncline axis, average annual temperature, elevation, aridity, and average annual rainfall play important roles during the predicting process of sub-model four.

5.6 Discussion

5.6.1 Feature Importance Analysis

The distance from syncline axis is at the dominant place among the 20 factors for the parent model, sub-model one, and sub-model four. Factors associated with water, such as average annual rainfall and the distance from rivers, are also important

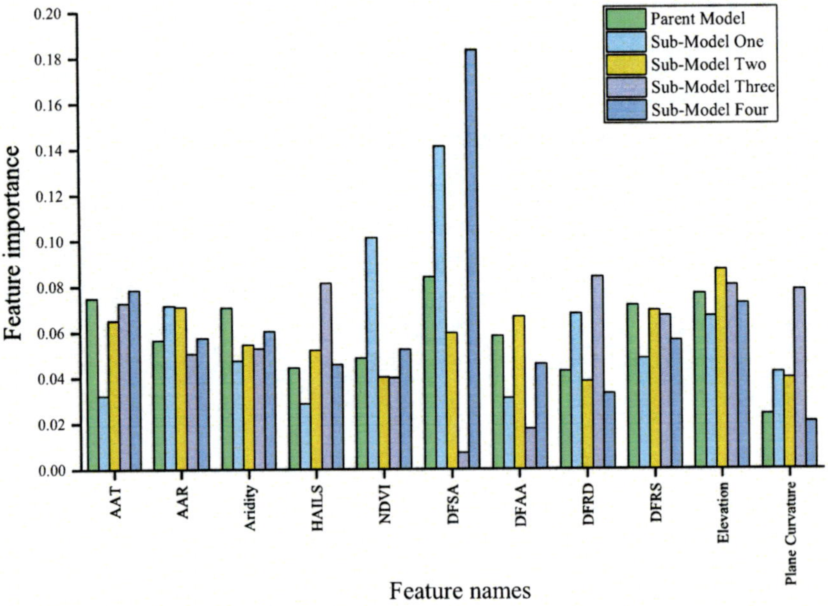

Fig. 5.6 Top five important factors of different models. AAT: average annual temperature; AAR: average annual rainfall; DFSA: distance from syncline axis; DFAA: distance from anticline axis; DFRD: distance from road; DFRS: distance from rivers

places. The triggering factors of shallow landslides are highly dependent on the rainfall water infiltration and its further redistribution (Ivanov et al. 2020). Due to its softening effect and long-term erosion effect, the distance from rivers has a significant influence on the development of landslides (Wang et al. 2017). Temperature has a remarkable effect on landslide formation; experimental results indicated that the shear strength of slip surface soils reduces with decreasing temperature, which will negatively affect slope instability (Shibasaki et al. 2016). Therefore, obtained by the accumulated temperature and rainfall, aridity is also of great importance. Although different plants generally have a different contribution of rainfall to soil water (Zhang et al. 2022a, b, c), which will affect slope stability differently; on average, the effect of water uptake from the plant cover makes the vegetated slopes averagely 12.84% drier, and matric suctions three times higher than the fallow slope, which contributes to slope stability (Gonzalez-Ollauri and Mickovski 2017); so NDVI is taken as one of the main considerations. Unlike vegetation cover, human-made land covers usually have negative influences on slope stability. The development of new building areas can potentially increase susceptibility to landslides (Promper et al. 2014). Road networks not only directly destabilize existing slopes by disturbing their original structures during construction processes, but transport activities also have negative effects on slope stability. Similarly, HAILS is another influential factor. For topological factors, the two most important factors, in this case, are elevation and plane curvature.

5.6.2 Model Comparison

When dealing with the whole region, the model tends to capture information that is more general, and thus, its feature importance has a higher universality. In our case, six major synclines are lying in the whole domain, and their generally dominant effects on identifying landslide/non-landslide points are successfully captured by the parent model. Aridity, distance from rivers, and average annual temperature affect the formation of landslides in meteorological and hydrological aspects. Although ranked fourth as the most important one, elevation is the most important topological factor in the research area.

After dividing the whole region into different sub-zones, the models will have the chance to learn the knowledge that is specifically right for each sub-zone. Thus, their performances will be improved.

Sub-model One. Located in the south of Yunyang County, sub-zone I occupies 741.7 km^2, which is only 20% of the entire research area. Nevertheless, it is crossed by two of the six major synclines, which makes the syncline effect more obvious, and, therefore, the distance from the syncline axis is still the most important factor for sub-zone I. Similarly, the landslide susceptibility in sub-zone I is also affected by both hydrological and topological factors, namely average annual rainfall and elevation. However, specifically, as it is crossed by the major highway S305, land cover (NDVI) and human activities (distance from road) are the other two main influential factors.

Sub-model Two. As we discussed before, sub-zone II is the most special sub-zone since it is crossed by the TGRA, most of the landslides in Yunyang County were impacted by water occurred here. Therefore, theoretically, the landslide formation in sub-zone II is sensitive to hydrological factors. In this case, the average annual rainfall and the distance from rivers are ranked as the second- and the third-most important factors. The reason why elevation is of the greatest importance is that lower places are usually prone to the influence of the periodic variation of reservoir water level. Statistically, 76% of the landslides impacted by water in sub-zone II occurred below the elevation of 249 m. The tectonic action and natural factors also contribute a lot to the landslide formation in sub-zone II; the distance from anticline axis and the average annual temperature are ranked in fourth and fifth place, respectively.

Sub-model Three. For sub-zone III, specifically, its data points are relatively far away from the major synclines, which mitigates the importance of the feature. Instead, as an area crossed by major highways and dense road networks, human activities, including distance from road and HAILS, are the two most important factors that have the main contribution to the landslide formation of the sub-zone. Topographical factors, namely elevation and plane curvature, are ranked in third and fourth place, respectively, and the average annual rainfall is evaluated as the fifth-most important factor.

Sub-model Four. Crossed by three major synclines, the landslide identification in sub-zone IV is also significantly dependent on the distance from the syncline axis. Furthermore, since it is part of the low hills section crossing Yunyang, Fengjie, and Kaizhou, the formation of landslides in this area is primarily influenced by the factors associated with the stability of mountain bodies, such as the average annual temperature, elevation, aridity, and the average annual rainfall.

5.7 Summary and Conclusion

In this chapter, Yunyang County is manually zoned into four parts based on the qualitative analysis of geological hazards exploration in Chongqing City, including the mechanism of landslide formation and sliding failure and geomorphic unit characteristics. Based on the qualitative analysis result, five random forest landslide susceptibility models are constructed using historical landslides data points and 20 relating factors for the following quantity analysis. These models, including a parent model and four sub-models, are optimized by the grid search method individually. A comparison between the parent model and the combination of the sub-models is conducted. The following conclusions are drawn:

The AUC value of the parent model achieves 0.872, which shows that the traditional RF with the hyper-parameters tuned by the grid search method has a reliable performance on landslide susceptibility mapping. In this study, synclines have the most important effects on the formation of landslides in Yunyang County, followed by aridity, elevation, distance from rivers, and average annual temperature.

However, more general information extracted from "mainstream" landslides would usually cover that of the "minority" landslides when treating a large region equally, resulting in low information utility and the inability to identify potential landslides under special geological conditions. With enough data points, experience-based zoning before modeling is proved to be an effective solution to the issue; the qualitative analysis serves the purpose of pre-classification based on the information from geological hazards exploration, which groups the landslides that occurred under similar geological conditions, and thus enables the models to obtain the specific knowledge under each condition. Therefore, in our case, while the traditional RF obtained the general prediction skill for the entire region of Yunyang County, all the sub-models have become "experts" in their respective sub-areas. The test AUC values of sub-model one to four are 8.8, 2.3, 1.9, and 9.1% higher than those of the parent model. Furthermore, the proposed method also contributes to further revealing the key factors that include local landslide instability under specific geological conditions, which can be used by planners and policymakers for a more specific and accurate landslide control in certain areas, thus further improving the safety of life and public property.

For sub-zone I, the top five conditioning factors are distance from syncline axis, NDVI, average annual rainfall, distance from road, and elevation. For sub-zone III, without the influence of major synclines, its top factors are distance from road, HAILS, elevation, plane curvature, and average annual temperature. For sub-zone IV, the distance from syncline axis becomes the most important factor again, and it is followed by average annual temperature, elevation, aridity, and average annual rainfall.

Sub-zone II is crossed by the Three Gorges Reservoir area. Suffered by periodic variation in reservoir water level and the impacts of other factors related to the reservoir band, the modified method based on general conditioning factors has relatively less effect on improving the accuracy of the mapping. The effect of more specific factors on the formation of landslides on the banks of the reservoir will be analyzed in further research. In the case of this chapter, the results of sub-model two point out that elevation, average annual rainfall, distance from rivers, distance from anticline axis, and average annual temperature are the top five conditioning factors among the existing twenty factors for sub-zone II.

References

Akar Ö, Güngör O (2015) Integrating multiple texture methods and NDVI to the random forest classification algorithm to detect tea and hazelnut plantation areas in northeast Turkey. Int J Remote Sens 36:442–464. https://doi.org/10.1080/01431161.2014.995276

Brabb E (1984) Innovative approaches to landslide hazard mapping. In: Proceedings of the 4th international symposium on landslides, Toronto, ON, Canada, vol 1, pp 307–324

Breiman L (1996) Bagging predictors. Mach Learn 24:123–140. https://doi.org/10.1007/bf00058655

Cutler A, Stevens JR (2006) Random forests for microarrays. Meth Enzymol 411:422–32. https://doi.org/10.1016/S0076-6879(06)11023-X

Feng Q, Liu J, Gong J (2015) Urban flood mapping based on unmanned aerial vehicle remote sensing and random forest classifier—a case of Yuyao China. Water 7:1437–1455. https://doi.org/10.3390/w7041437

Girma F, Raghuvanshi TK, Ayenew T, Hailemariam T (2015) Landslide hazard zonation in Ada Berga district, central Ethiopia—a GIS based statistical approach. J Geomat 9:25–38

Gonzalez-Ollauri A, Mickovski SB (2017) Hydrological effect of vegetation against rainfall-induced landslides. J Hydrol 549:374–387. https://doi.org/10.1016/j.jhydrol.2017.04.014

He Q, Wang M, Liu K (2021) Rapidly assessing earthquake-induced landslide susceptibility on a global scale using random forest. Geomorphology 391:107889. https://doi.org/10.1016/j.geomorph.2021.107889

Hengl T, Heuvelink GBM, Kempen B, Leenaars JGB, Walsh MG, Shepherd KD, Sila A, MacMillan RA, Mendes De Jesus J, Tamene L, et al (2015) Mapping soil properties of Africa at 250 m resolution: random forests significantly improve current predictions. PLoS ONE 10:e0125814. https://doi.org/10.1371/journal.pone.0125814

Hu Q, Zhou Y, Wang S, Wang F (2020) Machine learning and fractal theory models for landslide susceptibility mapping: case study from the Jinsha River Basin. Geomorphology 351:106975. https://doi.org/10.1016/j.geomorph.2019.106975

Huang F, Huang J, Jiang S, Zhou C (2017) Landslide displacement prediction based on multivariable chaotic model and extreme learning machine. Eng Geol 218:173–186. https://doi.org/10.1016/j.enggeo.2017.01.016

Huang F, Chen J, Liu W, Huang J, Hong H, Chen W (2022) Regional rainfall-induced landslide hazard warning based on landslide susceptibility mapping and a critical rainfall threshold. Geomorphology 408:108236. https://doi.org/10.1016/j.geomorph.2022.10823

Ivanov V, Arosio D, Tresoldi G, Hojat A, Zanzi L, Papini M, Longoni L (2020) Investigation on the role of water for the stability of shallow landslides—insights from experimental tests. Water 12:1203. https://doi.org/10.3390/w12041203

LaValle SM, Branicky MS (2004) On the relationship between classical grid search and probabilistic roadmaps. Springer Tracts in Advanced Robotics: Berlin/Heidelberg, Germany, pp 59–75. https://doi.org/10.1007/978-3-540-45058-0_5

Lee S, Pradhan B (2007) Landslide hazard mapping at Selangor, Malaysia using frequency ratio and logistic regression models. Landslides 4:33–41. https://doi.org/10.1007/s10346-006-0047-y

Lim CH, Kim HJ (2022) Can forest-related adaptive capacity reduce landslide risk attributable to climate change?—Case of Republic of Korea. Forests 13:49. https://doi.org/10.3390/f13010049

Lin L, Lin Q, Wang Y (2017) Landslide susceptibility mapping on a global scale using the method of logistic regression. Nat Hazards Earth Syst Sci 17:1411–1424. https://doi.org/10.5194/nhess-17-1411-2017

Liu CH, Chamberlain BP, Little DA, Cardoso Â (2017) Generalising random forest parameter optimisation to include stability and cost. Mach Learn Knowl Discov Databases 10536:102–113. https://doi.org/10.1007/978-3-319-71273-4_9

Lujan-Moreno GA, Howard PR, Rojas OG, Montgomery DC (2018) Design of experiments and response surface methodology to tune machine learning hyperparameters, with a random forest case-study. Expert Syst Appl 109:195–205. https://doi.org/10.1016/j.eswa.2018.05.024

Mathew J, Jha VK, Rawat GS (2008) Landslide susceptibility zonation mapping and its validation in part of Garhwal Lesser Himalaya, India, using binary logistic regression analysis and receiver operating characteristic curve method. Landslides 6:17–26. https://doi.org/10.1007/s10346-008-0138-z

Nakileza BR, Nedala S (2020) Topographic influence on landslides characteristics and implication for risk management in upper Manafwa catchment, Mt Elgon Uganda. Geoenviron Disast 7:27. https://doi.org/10.1186/s40677-020-00160-0

Nefeslioglu HA, Gokceoglu C, Sonmez H, Gorum T (2011) Medium-scale hazard mapping for shallow landslide initiation: the Buyukkoy catchment area (Cayeli, Rize, Turkey). Landslides 8:459–483. https://doi.org/10.1007/s10346-011-0267-7

Nhu V-H, Shirzadi A, Shahabi H, Singh SK, Al-Ansari N, Clague JJ, Jaafari A, Chen W, Miraki S, Dou J et al (2020) Shallow landslide susceptibility mapping: a comparison between logistic model tree, logistic regression, naïve Bayes tree, artificial neural network, and support vector machine algorithms. Int J Environ Res Public Health 17:2749. https://doi.org/10.3390/ijerph17082749

Petley D (2012) Global patterns of loss of life from landslides. Geology 40:927–930. https://doi.org/10.1130/g33217.1

Promper C, Puissant A, Malet J-P, Glade T (2014) Analysis of land cover changes in the past and the future as contribution to landslide risk scenarios. Appl Geogr 53:11–19. https://doi.org/10.1016/j.apgeog.2014.05.020

Raghuvanshi TK, Negassa L, Kala PM (2015) GIS based grid overlay method versus modeling approach—a comparative study for landslide hazard zonation (LHZ) in Meta Robi District of west showa zone in Ethiopia. Egypt J Remote Sens Space Sci 18:235–250. https://doi.org/10.1016/j.ejrs.2015.08.001

Reichenbach P, Rossi M, Malamud BD, Mihir M, Guzzetti F (2018) A review of statistically-based landslide susceptibility models. Earth-Sci Rev 180:60–91. https://doi.org/10.1016/j.earscirev.2018.03.001

Sheng L, Wang W, Zhu W (2016) China statistical yearbook 2016. China Statistics Press, Beijing, China (In Chinese)

Shibasaki T, Matsuura S, Okamoto T (2016) Experimental evidence for shallow, slow-moving landslides activated by a decrease in ground temperature. Geophys Res Lett 43:6975–6984. https://doi.org/10.1002/2016gl069604

Smith HG, Spiekermann R, Betts H, Neverman AJ (2021) Comparing methods of landslide data acquisition and susceptibility modelling: examples from New Zealand. Geomorphology 381:107660. https://doi.org/10.1016/j.geomorph.2021.107660

Sun D, Wen H, Wang D, Xu J (2020) A random forest model of landslide susceptibility mapping based on hyperparameter optimization using Bayes algorithm. Geomorphology 362:107201. https://doi.org/10.1016/j.geomorph.2020.107201

Sun D, Xu J, Wen H, Wang D (2021) Assessment of landslide susceptibility mapping based on Bayesian hyperparameter optimization: a comparison between logistic regression and random forest. Eng Geol 281:105972. https://doi.org/10.1016/j.enggeo.2020.105972

Sun D, Gu Q, Wen H, Shi S, Mi C, Zhang F (2022) A hybrid landslide warning model coupling susceptibility zoning and precipitation. Forests 13:827. https://doi.org/10.3390/f13060827

Taalab K, Cheng T, Zhang Y (2018) Mapping landslide susceptibility and types using random forest. Big Earth Data 2:159–178. https://doi.org/10.1080/20964471.2018.1472392

Wang F, Xu P, Wang C, Wang N, Jiang N (2017) Application of a GIS-based slope unit method for landslide susceptibility mapping along the Longzi River, southeastern Tibetan Plateau, China. ISPRS Int J Geo-Inf 6:172. https://doi.org/10.3390/ijgi6060172

Wang L, Zhang Z, Huang B, Hu M, Zhang C (2021a) Triggering mechanism and possible evolution process of the ancient Qingshi landslide in the Three Gorges Reservoir. Geomat. Nat. Hazards Risk 12:3160–3174. https://doi.org/10.1080/19475705.2021.1998230

Wang Y, Tang H, Huang J, Wen T, Ma J, Zhang J (2022) A comparative study of different machine learning methods for reservoir landslide displacement prediction. Eng Geol 298:106544. https://doi.org/10.1016/j.enggeo.2022.106544

Wang X, Li S, Liu H, Liu L, Liu Y, Zeng S, Tang Q (2021b) Landslide susceptibility assessment in Wenchuan County after the 5.12 magnitude earthquake. Bull Eng Geol Environ 80:5369–5390. https://doi.org/10.1007/s10064-021-02280-9

Wu C, Hong L, Wang L, Zhang R, Pijush S, Zhang W (2022) Prediction of wall deflection induced by braced excavation in spatially variable soils via convolutional neural network. Gondwana Res. https://doi.org/10.1016/j.gr.2022.06.011

Yesilnacar E, Topal T (2005) Landslide susceptibility mapping: a comparison of logistic regression and neural networks methods in a medium scale study, Hendek Region (Turkey). Eng Geol 79:251–266. https://doi.org/10.1016/j.enggeo.2005.02.002

Yin Y, Wang L, Zhang W, Dai Z (2022) Research on the collapse process of a thick-layer dangerous rock on the reservoir bank. Bull Eng Geol Environ 81:109. https://doi.org/10.1007/s10064-022-02618-x

Zhang W, Liu S, Wang L, Samui P, Chwała M, He Y (2022b) Landslide susceptibility research combining qualitative analysis and quantitative evaluation: a case study of Yunyang County in Chongqing, China. Forests 13:1055. https://doi.org/10.3390/f13071055

Zhang Y, Zhang B, Xu Q, Gao D, Xu W, Ren R, Jiang J, Wang S (2022c) The Effects of Plant and Soil Characteristics on Partitioning Different Rainfalls to Soil in a Subtropical Chinese Fir Forest Ecosystem. Forests 13:123. https://doi.org/10.3390/f13010123

Zhang W, Wu C, Zhong H, Li Y, Wang L (2020) Prediction of undrained shear strength using extreme gradient boosting and random forest based on Bayesian optimization. Geosci Front 12:469–477. https://doi.org/10.1016/j.gsf.2020.03.007

Zhang W, Li H, Han L, Chen L, Wang L (2022a) Slope stability prediction using ensemble learning techniques: a case study in Yunyang County, Chongqing, China. J Rock Mech Geotech Eng. https://doi.org/10.1016/j.jrmge.2021.12.011

Zhou X, Wen H, Zhang Y, Xu J, Zhang W (2021) Landslide susceptibility mapping using hybrid random forest with GeoDetector and RFE for factor optimization. Geosci Front 12:101211. https://doi.org/10.1016/j.gsf.2021.101211

Zhou X, Wen H, Li Z, Zhang H, Zhang W (2022) An interpretable model for the susceptibility of rainfall-induced shallow landslides based on SHAP and XGBoost. Geocarto Int TGEI 1–27. https://doi.org/10.1080/10106049.2022.2076928

Chapter 6
Application of Transfer Learning to Improve Landslide Susceptibility Modeling Performance

6.1 Introduction

With climate change, the frequencies of specific extreme weather and climate events such as floods and rainstorms have increased a lot (Stott 2016), which subsequently accelerate the surge of landslides in the whole world. Being regarded as one of the most prevalent and destroying geological hazards (Ji et al. 2022), landslides can potentially cause severe damage to infrastructures and houses and sometimes even lead to loss of life. Therefore, reliable regional landslide susceptibility assessment has become increasingly important for planners and researchers to mitigate such social-eco impacts (Tonini et al. 2020). Landslide susceptibility assessment is obtained from quantitative and qualitative analysis of the local terrain conditions and historical landslide distributions, describing the likelihood of landslide occurrence in a certain area (Brabb 1984). It can be used as a reference of the potential distribution of future landslides based on the assumption that landslides will occur under similar conditions that produced historical landslides (Guzzetti et al. 2006). The mainstream assessment methods can be classified into three basic types: knowledge-based methods, physical methods, and data-driven methods (Corominas et al. 2013). With the development of computational devices and artificial intelligence techniques, the application of data-driven methods (e.g., machine learning and deep learning-based methods) in landslide susceptibility has become a hot spot considering their desired capability of solving complex nonlinear problems.

Various machine learning-based methods were applied by previous researchers on landslide susceptibility assessment. Zhang et al. (2022) developed an ensemble learning-based method to predict the slop stability based on RF and XGBoost. Wang et al. (2021a) proposed a novel AI-powered object-based landslide susceptibility assessment method, which outperformed the other six AI algorithms based on traditional cell-based method. Zhao et al. (2019) applied LR method on assessing the landslide susceptibility in the northern Yueqing of Zhejiang Province, China. Goyes-Penafiel and Hernandez-Rojas (2021) assessed the landslide susceptibility index of Popaya, Colombia, based on the presented integration of LR and weights of evidence.

Z. Wengang et al., *Application of Machine Learning in Slope Stability Assessment*,
https://doi.org/10.1007/978-981-99-2756-2_6

Huang et al. (2017) proposed a self-organizing map network-based extreme learning machine model to calculate the landslide susceptibility indexes for Wanzhou district in the TGRA. By using the traditional machine learning-based methods above, sufficient governmental data and complete landslide inventories are the pre-requirements for obtaining efficient and credible landslide susceptibility mapping (Wang et al. 2021a). Although the fast development of earth observation equipment and analyzing software has made the data collecting processes easier (Wang et al. 2021b), due to various reasons such as unequal development of technology and low cost–benefit ratio of regional landslides analysis (Van Westen et al. 2006), compiling complete landslide inventory remains a challenge for some areas. Traditionally, the results of the landslide susceptibility assessment obtained from rare landslide records do not have enough reference value, and it remains until the emergence of advanced big data processing technics and transfer learning (TL) algorithms.

By transferring the knowledge from the data-rich region (i.e., source domain) to a new region with a relatively poor dataset (i.e., target domain) through model tasks, transfer learning is capable to effectively solve the problems caused by data deficiency (De Lima and Marfurt 2019). Previous studies mainly focused on evaluating the landslide susceptibility of specific regions by utilizing traditional machine learning-based methods. While their achievement and models are usually not suitable to the other regions and cannot be upgraded sustainably, transfer learning-based methods can accumulate the data from different regions and then transfer the obtained knowledge to a specific area, which cannot only dramatically extend the application of regional-specific landslide susceptibility models, but can also be continuously upgraded with more data input. With these merits, transfer learning-based methods are becoming increasingly popular in various fields with scarce datasets (Duan et al. 2018; Fang et al. 2013; Wei et al. 2014). However, it has not been popularized in the field of landslide susceptibility assessment.

In this chapter, we accomplished the first transfer learning application to landslides susceptibility based on a nationwide dataset and proposed a 1D CNN-bidirectional long short-term memory model on the basis of LandslideNet (Wang et al. 2021a) to deal with the input shaped as a one-dimensional array. It was used to extract the characteristics of the areas prone to landslides based on an area with dense data points (source domain) first, then the obtained knowledge was transferred to Chongqing for local landslide susceptibility analysis. In addition, we trained the TL model with 1, 5, and 10% of the obtained historical landslide points in Chongqing to simulate the data-limited cases. Then, the performance of traditional DL model and that of the TL model was compared to validate the following two hypotheses:

(1) TL models will significantly improve the accuracy of landslide susceptibility mapping in data-limited areas.
(2) TL models can also be used to further optimize the landslide susceptibility maps in data-rich areas.

6.2 Study Area

China is one of the countries with the densest geological hazard occurrence in the world (Wen et al. 2004). Because of its high population density (the second place in the world), the geological hazards-related injuries and deaths were always at a high stage. While the average geological hazards-related deaths in the USA is 25 people per year, according to China Statistical Yearbook, 10,966 people were killed by the landslides occurred from 2000 to 2015, which is approximately 690 landslides-related deaths per year in this country (Sheng et al. 2016). Therefore, more specific and precise landslide susceptibility mapping is needed for planners and policymakers to avoid the serious outcomes caused by large- and median-scale landslides and mitigate the potential dangers of small ones. However, 94% of the geological hazards from 2015 to 2020 occurred in rural areas where there was usually little financial support for obtaining complete landslides inventory maps. As discussed before, it is difficult to propose a precise landslide susceptibility map by only using the incomplete inventory, and the proposed 1D TL model in this study is aimed to solve this problem.

Belonging to the TGRA, Chongqing City (Spans from 105° 11′ E to 110° 11′ E and 28° 10′ N to 32° 13′ N) is the economic, financial, technological innovation, shipping, and trade logistics center of the southwest China. It is located in a mountainous area, where 92% of its total area is occupied by mountains and hills, making it also known as "mountain city". The elevation of Chongqing City is between −2 and 2782 m, and it is crossed by five major rivers, namely the Yangtze River, Jialing River, Wu River, Qi River, and Fu River (Fig. 6.1). The mountainous condition combined with various landslide-promoting factors such as developed surface water networks, strongly cut terrain, complex rock and soil structure and geological structure makes this area seriously prone to geological hazards (Yang et al. 2022; Li et al. 2020; Nian et al. 2020). Meanwhile, after the construction of the Three Gorges Reservoir, the water storage process started the periodic variation of reservoir water level, which would potentially accelerate the formation of local geological hazards (Yang et al. 2019). Meanwhile, under the effect of sub-tropical monsoon humid climate, Chongqing City has an annually mean precipitation from 1002 to 1671 mm, whose peak is about twice of the global average precipitation (876 mm/year (Shen and Tafolla 2014). The large amount of rainfall interacted with periodic variation of reservoir water level, further increasing the landslides susceptibility of this area (Yan et al. 2019).

However, as the most populous city in China and the second in the world (more than thirty-two million residents in the year of 2022), dense occurrences of geological hazards caused serious potential dangers to local residents, which significantly affect the safety of public health and property. According to the analysis and statistics of the local government, there are currently 14,926 geological hazard-prone points in Chongqing City, most of which are distributing along water region. The coordinates of 11,142 historical landslide points in Chongqing City were obtained from the data collected by the Chinese Academy of Sciences.

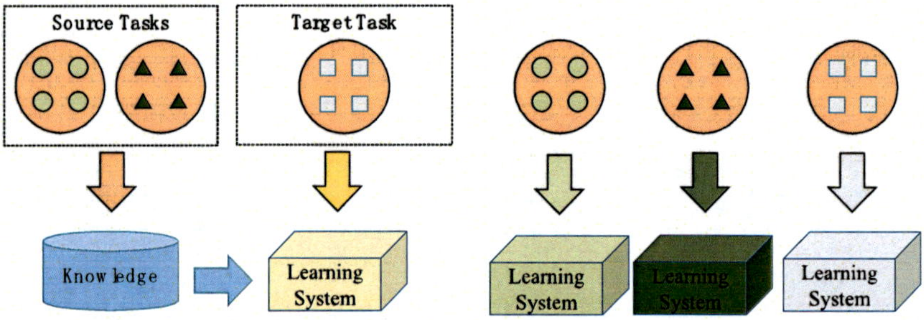

Fig. 6.1 Transfer learning

6.3 Transfer Learning

The modern mission of transfer learning was defined as recognizing and applying the knowledge and skills learned in previous tasks to novel tasks (Fig. 6.1) by the Broad Agency Announcement of Defense Advanced Research Projects Agency's Information Processing Technology Office in 2005. Unlike traditional machine learning which is basically constructed by one dataset $D = \{(x_1, y_1), \ldots, (x_i, y_i)\}$, where $x_i \in X_i$ is the input and $y_i \in Y_i$ is the corresponding output, the establishment of a TL model needs two different datasets. They are defined as source domain data $D_S = \{(x_{S_1}, y_{S_1}), \ldots, (x_{S_{nS}}, y_{S_{nS}})\}$ and target domain data $D_T = \{(x_{T_1}, y_{T_1}), \ldots, (x_{T_{nT}}, y_T)\}$, where $x_{S_i} \in X_s$ and $x_{T_i} \in X_T$ are the data instance for source domain and target domain, respectively, and $y_{S_i} \in y_s$ and $y_{T_i} \in y_T$ are the corresponding class labels. Given a source domain D_S and a learning task T_S, a target domain D_T and learning task T_T, transfer learning aims to help improve the learning of the target predictive function $f_T(\cdot)$ in D_T using the knowledge in D_S and T_S, where $D_S \neq D_T$ or $T_S \neq T_T$.

Generally, a transfer learning model is to pre-train a base network based on the source domain dataset, then the parameters of the first several layer(s) will be kept, and the other remaining layers will be retrained by using the target domain dataset. The errors will be backpropagated from the new task into the base features to fine-tune the parameters of the layers that were not frozen, making the new model more suitable for the target domain (Vrbancic and Podgorelec 2020). Therefore, transfer learning is capable to utilize both "outdated but abundant knowledge" and "fashionable but small amount of knowledge", which can theoretically better accomplish some tough tasks. Shao et al. (2019) developed a novel deep transfer learning framework to achieve fast and highly accurate machine fault diagnosis. Gao et al. (2018) applied transferring learning approach to the new ImageNet, namely Structural ImageNet for structural damage recognition by using limited number of labeled images. Pathak et al. (2022) applied a transfer learning-based classification model for COVID-19 disease detection.

The properties of transfer learning are perfect for landslide susceptibility analysis, which can fully utilize its knowledge-sharing features to overcome the difficulties of completely landslide inventorying and thus reduce the workload of landslide susceptibility mapping. With transfer learning-based methods, the spatial restriction of landslide points will be mitigated, and the knowledge obtained from the other areas can be transferred to the target region for precise landslide susceptibility mapping, especially for those regions with limited data, such as newly developed regions or undeveloped urban areas. In order to verify the hypotheses discussed at the end of the introduction section, the first countrywide application of transfer learning-based methods on landslide susceptibility is proposed by this study.

6.4 Methodology

6.4.1 Data Preparation

Data preparation plays a vital role in machine learning applications, and the quality of data usually determines the performance of the developed model. In this study, the coordinates of 108,705 historical landslide points in the mainland of China were collected from the Chinese Academy of Sciences. The ratios of the recorded landslide points in different regions were displayed in Table 6.1. Possibly missing records might exist due to the low population density in the north of China, and 86.64% of the recorded historical landslides (94,191/108,705) were located at part of the southern the mainland of China, which is defined as the area with dense landslides in this case. In order to mitigate the problem of low reliability caused by potentially missing records, we decided to only use the area with dense landslides (except Chongqing City) as our source domain, where there are 83,048 recorded historical landslide points (positive samples). Then, as suggested by Mathew et al. (2008), we randomly extracted the same number of non-landslide points from landslide-free areas as negative samples. Similarly, as the target domain, the 11,147 recorded historical landslide points of Chongqing City were treated as the positive samples of the target domain dataset, and in the same way, 11,147 negative samples in target domain were obtained (Table 6.2).

The mechanism of landslide occurrence is a complex topic that is associated with the joint effects of various natural factors and human activities (Sun et al. 2021a). Therefore, in order to achieve as accurate prediction as possible, abundant kinds of conditioning factors are needed. Reichenbach et al. (2018) found that there were total

Table 6.1 Ratios of historical landslide records in different regions

Region	Number of landslides	Percentage (%)
The mainland of China	108,705	100
Area with dense landslides	94,191	86.64
Chongqing City	11,147	10.25

Table 6.2 Sample numbers in different domains

Domain	Number of landslide points (positive samples)	Number of non-landslide points (negative samples)
Source domain	83,048	83,048
Target domain	11,147	11,147

596 distinct factors controlling landslide formation after reviewing landslide sensitivity evaluation-related studies from 1983 to 2016. However, it is laborious and time-consuming to take them all into consideration. According to previous studies about AI-based landslide susceptibility analysis, topographic, hydrological, geological, land cover, natural, and human-related factors were generally used for model training (Nhu et al. 2020; Zhou et al. 2021). Therefore, we planned to use natural factors such as annually mean temperature and annually mean precipitation; topographical factors (aspect, profile curvature, plane curvature, slope, elevation, and RDLS); human-related factors, namely distance from road, population, and GDP; land cover types (land use, NDVI); hydrological factors including aridity, IM, TWI, and distance from water; and geological factors (lithology, soil type, and distance from faults) for landslide susceptibility modeling in this study.

Most of the conditioning features were obtained from the Resource and Environment Science and Data Center. In order to get more information from the raw data, GIS was applied, by which the fishnet of the study area was created first, then the near function of GIS was used based on the created fishnet and corresponding vector datasets to generate and rasterize the features associated with distances such as distance from water, distance from roads, and distance from faults. In addition, aspect, profile curvature, plane curvature, slope, RDLS, and TWI were obtained by GIS processing of DEM. Thus, both the raw data and the secondary data were obtained, and their resolutions and sources were arranged in Table 6.3.

6.4.2 Data Extraction and Model Preparation

The workflow of the transfer learning-based landslide susceptibility analysis is summarized in Fig. 6.2. The data layers and sample points in the whole domain were separated into source domain and target domain. The source domain is defined as the knowledge-obtaining region, usually consisting of data-rich sources. With no limitation for the number of inputs, it has a high level of flexibility and expandability. The volume of the dataset can be expanded when any additional data points are available, which means those data points can be simply added into the source domain dataset for more knowledge extraction. In this case, the source domain is the area with dense landslides which was highlighted in Fig. 6.2, and its domain can be expanded whenever new reliable samples were obtained in the future work. The

Table 6.3 Conditioning factors

Feature ID	Feature name	Revolution (m)	Data sources
1	Annually mean temperature	500	Resource and environment science and data center
2	Annually mean rainfall	500	Resource and environment science and data center
3	GDP	1000	Resource and environment science and data center
4	Distance from road	30	Resource and environment science and data center + near function of GIS
5	Population	1000	Resource and environment science and data center
6	Land use	30	Resource and environment science and data center
7	NDVI	30	Resource and environment science and data center
8	Aridity	500	Resource and environment science and data center
9	IM	500	Resource and environment science and data center
10	TWI	30	GIS processing of DEM
11	Distance from water	30	Resource and environment science and data center + near function of GIS
12	Distance from faults	30	Resource and environment science and data center + near function of GIS
13	Soil type	30	Resource and environment science and data center
14	Lithology	30	Resource and environment science and data center
15	Elevation	30	Resource and environment science and data center
16	Slope	30	GIS processing of DEM
17	Planform curvature	30	GIS processing of DEM
18	Profile curvature	30	GIS processing of DEM
19	Aspect	30	GIS processing of DEM
20	RDLS	30	GIS processing of DEM

conditioning factors were extracted to both the positive and negative sample points in the source domain for constructing the source domain dataset D_S. Then, 80% of them were randomly selected to pre-train the proposed deep learning model, and the rest of them served for testing purposed. After obtaining the well-trained landslide susceptibility analysis model, the extracted knowledge can be applied to predict the landslide susceptibility of the target domain.

As the knowledge-application region, target domain is composed by limited data which is usually incomplete for training a reliable model. Unlike source domain, the target domain for a specific task is fixed. In this case, the target domain is Chongqing City. The target domain dataset D_{T} was obtained in the same way as that for the source domain dataset, where $D_{\mathrm{T}} \neq D_{\mathrm{S}}$. We then used 1, 5, and 10% of the data points in the dataset to simulate incomplete landslide inventory, which is generally

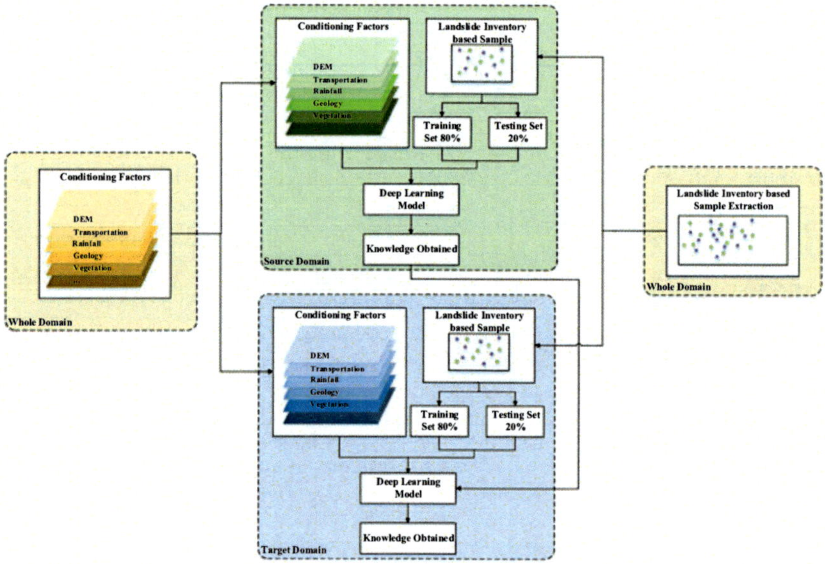

Fig. 6.2 Workflow of transfer learning-based landslide susceptibility analysis

not sufficient to produce a reliable model for landslide susceptibility analysis. But in this study, the knowledge obtained from source domain can be used to solve such issue. As the target task is quite similar to the source task, the knowledge contained in the pre-trained model is also suitable for predicting the landslide susceptibility of the target domain. With transfer learning technology, the pre-trained model can be applied to the target domain dataset for more specific knowledge extraction and thus generate the landslide susceptibility map for the target domain. During transfer learning process, the architecture of the pre-trained model was kept, and none of the parameters was reinitialized. In order to make full use of the stored knowledge, some of the layers were frozen, whose parameters were kept the same from the beginning to the end of the transfer learning process. Part of the layers were trainable, which could be further optimized by the limited data in the target domain, and thus generated precise landslide susceptibility prediction for the target domain.

The architecture of LandslideNet (Wang et al. 2021a) displaced an extreme performance on landslide susceptibility mapping, which was originally used to deal with 2D data obtained by the object-based approach. In this case, we adjusted the model to make it suit for the one-dimensional inputs by basically switching the convolution kernel from 2D into 1D. As shown in Fig. 6.3, the 1D LandslideNet is composed by seven modules. The first four of them are convolutional modules (Conv 1 to Conv 4), followed by two BiLSTM modules (BiLSTM1 and BiLSTM2), and one fully connected module which contains three fully connected layers. Issues associated with overfitting have always existed in deep learning-based models, which harms the robust performance of the models to a very large degree (Rice et al. 2020). To mitigate overfitting, dropout layers were applied in both convolutional modules and BiLSTM models.

In addition, because of the much larger spatial domain and the volume of collected data, we proposed a deeper network by trial and error, namely 1D LandslideNetPro. As displayed in Fig. 6.4, it is composed by nine modules, where five of them are convolutional modules (Conv 1 to Conv 5). Among them, Conv 1 contains three convolutional layers, Conv 2 to Conv 4 are composed by two convolutional layers, and Conv 5 has only one convolutional layer. Also, in order to avoid the low training efficiency caused by internal covariate shift and the low convergence efficiency caused by gradient vanishing and gradient explosion (Ioffe and Szegedy 2015),

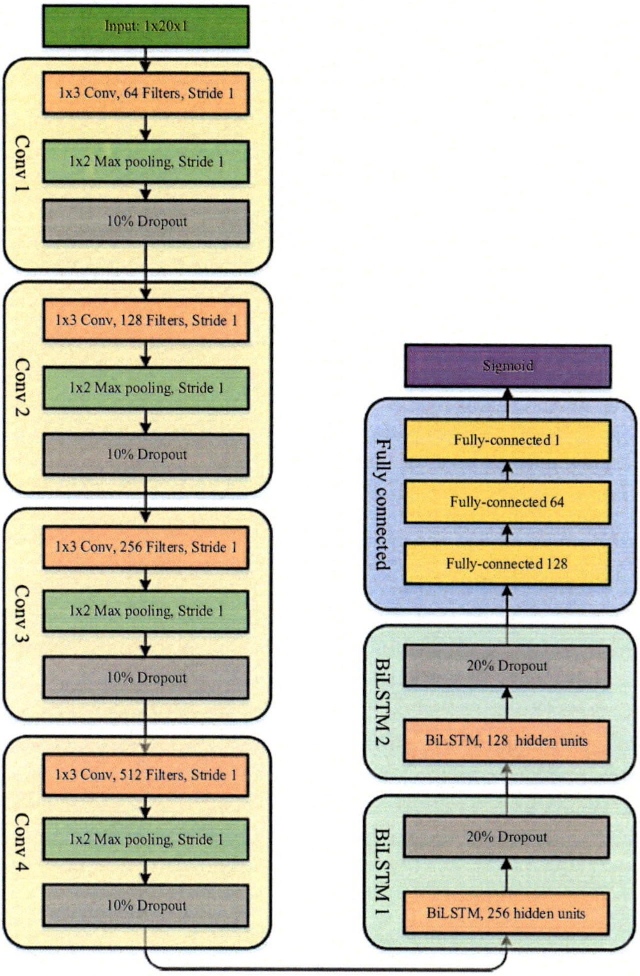

Fig. 6.3 Architecture of 1D LandslideNet

we added BanchNormolization layers into all of the five convolutional modules. Following that, there are three BiLSTM modules and one fully connected module which is composed by four fully connected layers. Moreover, as an important factor, an appropriate learning rate is crucial for improving model performance. Generally, the weights of the model were randomly initialized at the start of training process at the start of training process, which might cause the model instable if the learning rate is at a relatively large level. Therefore, the warmup strategy was applied to make the model be trained with a low learning rate initially. It will then continuously increase until the model turns to stable to improve the learning efficiency. At a certain stage when the loss is about to reach the global minima, the learning rate should be low enough again to make the model reach the point as much as possible. With the ability of gradual reduction, cosine annealing can be used to deal with this issue effectively. Therefore, warmup-cosine annealing (WCA) learning rate adjusting strategy was adopted in this study.

6.4.3 Model Application and Evaluation

In the first step, the proposed model LandslideNetPro and the basic model LandslideNet with/without WCA learning rate adjusting strategy were all applied to the target domain dataset and source domain dataset for the performance comparison. Then the one with higher suitability for the study case was chosen as the base architecture of transfer learning for the further analysis. To investigate the feasibility of the proposed transfer learning-based method in both the data-limited case and the data-rich case, we varied the amount of training samples from 1 to 95% during model training process. As displayed in Table 6.4, there are totally twelve modeling scenarios with different ratios of training samples (i.e., 1, 5, 10, 20, 30, 40, 50, 60, 70, 80, 90, and 95%), and the rest of them were used for testing purposes. To eliminate the influence of the magnitude of different conditioning features on modeling, normalization is always applied before model training. In this case, z-score normalization was adopted (Eq. 6.1) for each feature. For a specific conditioning feature, where x_i is ith value, $\overline{X}$ represents the mean value of the feature, and std(X) means the standard variation of the feature.

In each scenario, the basic form of the outperformed model (which is referred to as the DL model in the following content) was trained with the specific ratio of the normalized training samples in the target domain dataset. Its transfer learning form (which is referred to as the TL model in the following content) was trained with the same target domain samples and the proposed transfer learning approach. Specifically, in order to make full use of the knowledge obtained from source domain, the training samples for one of the TL models were normalized by the mean and standard variation of the source domain dataset. Their prediction performances were then compared, in which their accuracy and loss were analyzed.

For a binary classification task, traditional index "accuracy" sometimes cannot represent its true prediction performance, especially at the situation dealing with

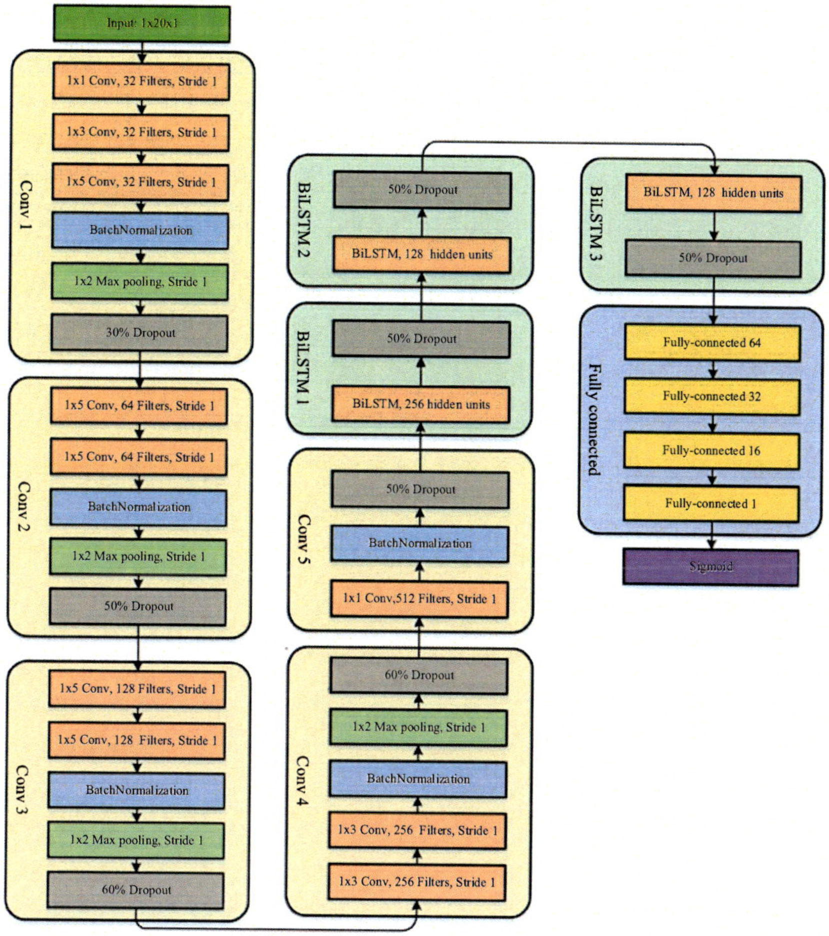

Fig. 6.4 Architecture of 1D LandslideNetPro

unbalanced dataset. For an extreme example, in a dataset which contains 9999 positive samples and only 1 negative sample, what the model learned might possibly be just to classify all of them into positive, the accuracy in this case is 99.99% according to the accuracy formula (Eq. 6.1). However, such a high accuracy does not mean that the model can really perform well in the real world, as it literally has no classification function at all. Therefore, the ROC curve was introduced, which represents the corresponding true positive rate $\left(\frac{TP}{TP+FN}\right)$ versus negative rate $\left(\frac{FP}{FP+TN}\right)$ at different cutoff values. The AUC is then used as an index representing the prediction performance of the model, and a higher AUC value indicates that the model has a better prediction performance.

Table 6.4 Modeling scenarios

Scenario number	Training samples percentage (extract number)	Test samples percentage (extract number)
1	1% (1661)	99% (164,435)
2	5% (8305)	95% (157,791)
3	10% (16,610)	90% (149,486)
4	20% (33,219)	80% (132,877)
5	30% (49,829)	70% (116,267)
6	40% (66,438)	60% (99,658)
7	50% (83,048)	50% (83,048)
8	60% (99,658)	40% (66,438)
9	70% (116,267)	30% (49,829)
10	80% (132,877)	20% (33,219)
11	90% (149,486)	10% (16,610)
12	95% (157,791)	5% (8305)

$$\text{Accuracy} = (\text{TP} + \text{TN})/(\text{TP} + \text{FP} + \text{TN} + \text{FN}) \quad (6.1)$$

where TP represents the true positive prediction, meaning both true and predicted classes are positive; similarly, TN stands for true negative prediction, implying both true and predicted classes are negative; on the other hand, FP means false positive prediction, referring the predicted classes are positive, but the true classes are negative; FN is called as false negative, which means the predicted classes are negative, while the true classes are positive.

The other index we used to quantify the model performance is logarithmic loss (Eq. 6.2):

$$\text{Logarithmic loss} = -\frac{1}{N}\sum_{i=1}^{N} y_i \log(\hat{y}_i) + (1 - y_i)\log(1 - \hat{y}_i), \quad (6.2)$$

where N represents the size of the dataset, y_i is the true class of ith sample (0 for negative and 1 for positive), and $\hat{y}_i$ is the ith output of the model, which implies the probability of ith sample being positive based on prediction. A lower logarithmic loss generally means better model performance.

6.5 Results and Discussion

For the comparative study, we randomly selected 80% of the samples from source domain dataset and target domain dataset to train the four models, respectively (i.e., 1D LandslideNet, 1D LandslideNet with WCA learning rate adjusting, 1D LandslideNetPro, and 1D LandslideNetPro with WCA learning rate adjusting). The AUC values and losses of the four models in both domains were shown in Fig. 6.5. From Fig. 6.5a–c, the AUC values gradually increase, and the losses gradually decrease with each upgrade, proving the effectiveness of the optimization. But in Fig. 6.5d, the loss of 1D LandslideNet with WCA learning rate adjusting is higher than that of 1D LandslideNet when it actually has a higher AUC value. It is a reasonable phenomenon from the definition of loss and AUC; as loss measures the output probability of a certain simple being positive class, it happens that the predicted classes are right, while the gaps between the output values and true classes (0 and 1) are relatively large. For example, when the best threshold was 0.5, the model outputted 0.4 for most of the negative samples and 0.6 for most of the positive samples, and both the AUC values and losses would be high.

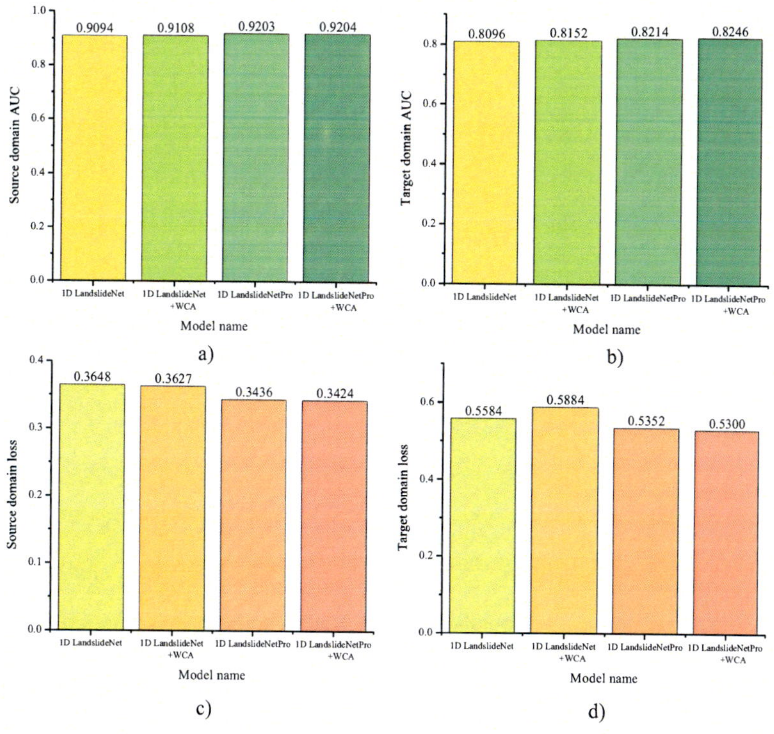

Fig. 6.5 AUC and loss of different models

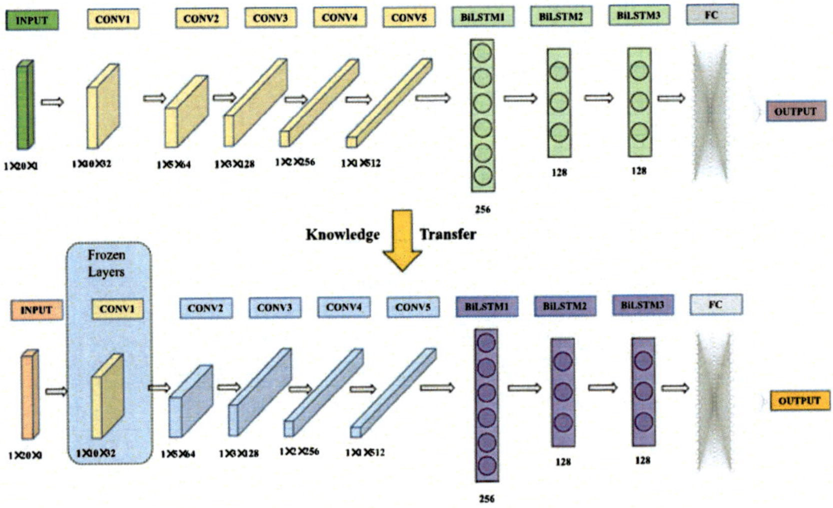

Fig. 6.6 Architecture of the proposed transfer learning model

Therefore, the 1D LandslideNetPro with WCA learning rate adjusting strategy was chosen for this study. Firstly, the DL model was directly established based on the aforementioned twelve different ratios of the target domain data. At the same time, the TL model (Fig. 6.6) was pre-trained by 80% of the random samples in source domain dataset for knowledge extraction, and the trained model was then applied to the twelve scenarios for transfer learning. To maximize the trainable parameters of the model, only the Conv 1 module is frozen during the transferring process.

The AUC values of different models and their improvements were displayed in Fig. 6.7. When the training data is limited (i.e., from 1 to 10% in this case), both the two TL models with different normalization standards achieved much higher AUC values for test dataset. In the case of 1% training samples, the gap between the AUC values of DL and TL models reached 0.206, which is an improvement of 41.1% based on Eq. (6.3). As for the TL model based on source domain normalization, the gap is even more obvious, reaching 0.256, corresponding to a 51.0% improvement. Similarly, in the cases of 5 and 10% training samples, the increases of the AUC values are 0.13 and 0.013, equivalent to 20.7% and 1.7% improvement, respectively. When the training samples were normalized based on the parameters in source domain datasets, the gaps increased to 0.15 and 0.016, corresponding to the improvements of 24.4 and 2.15%. With the increase of training sample size, such improvement becomes much smaller (fluctuating around 0.5%), but as the aforementioned hypothesis suggests, the TL model always outperforms the DL model even when the size of training samples reaches 95%, proving the effectiveness of the knowledge obtained from source domain. However, when the training samples were normalized based on the parameters of source domain dataset, the improvement starts to be less than the basic TL model after the training samples reach 20%, and it even produces negative effects when the size of training samples continually increases. This phenomenon suggests that when there are only limited training samples, normalizing them by the parameters obtained from source domain dataset is helpful for accuracy improvement, but when there are sufficient training samples, the parameters obtained from source domain dataset would be inappropriate for target domain dataset normalization.

$$\text{AUC improvement} = \frac{(\text{AUC}_{\text{TL}} - \text{AUC}_{\text{DL}})}{\text{AUC}_{\text{DL}}} \times 100\%. \tag{6.3}$$

The losses of different models in distinct scenarios were presented in Fig. 6.8. Similarly, when the training data is limited, both of the TL models have remarkable effects on loss reduction. According to Eq. (6.4), in the case of 1% training samples, the loss reduction of the basic TL model reached 8.8% and that of the other TL model achieved 16.1%. Similarly, in the case of 5%, 10%, and 20% training samples, the loss reduction of the basic TL model was 24.6%, 15.0%, and 11.3%, while that of the TL model with the samples normalized based on the parameters of source domain was 27.6%, 15.4%, and 11.4%, respectively. Such reduction effects gradually decreased as the size of training samples increased, but the basic TL model still outperforms the DL model under all scenarios.

$$\text{Loss reduction} = \frac{(\text{Loss}_{\text{DL}} - \text{Loss}_{\text{TL}})}{\text{Loss}_{\text{DL}}} \times 100\%. \tag{6.4}$$

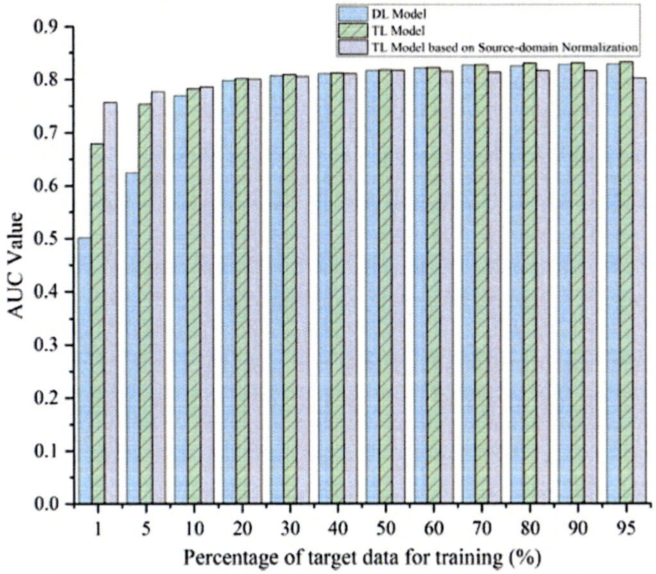

Fig. 6.7 AUC comparison between DL model and proposed TL models

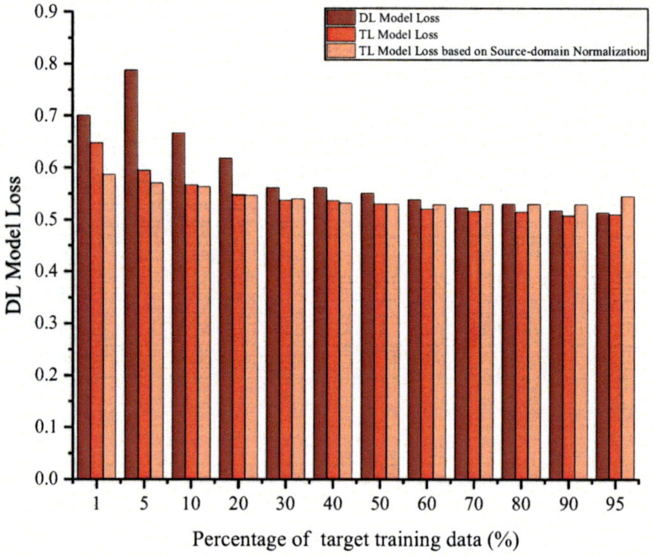

Fig. 6.8 Loss comparison between DL model and proposed TL models

The results above proved the first hypothesis that TL learning-based methods can significantly improve the accuracy of landslide susceptibility mapping in data-poor areas. Therefore, after comprehensively analyzing the results above, the TL model with the samples normalized based on the parameters of source domain was adopted in data-limited cases, and the basic TL model was suggested to be applied for improving the prediction accuracy in data-rich cases. The study area is divided into five categories with different levels of landslide susceptibility (very low, low, moderate, high, and very high) by the method of natural breaks based on the predicted susceptibility values. The DL model tends to overestimate the susceptibility when there is extremely limited data available, and thus it failed to provide reasonable landslide susceptibility maps in the cases when only 1% and 5% sample points were available for training, in which almost all of the areas were classified as very high susceptibility zones. The situation became better when the ratio increased to 10%. In comparison, the two TL models have always presented much more logical results even in the case where there were only 1% sample points for training. Especially for the TL model trained by the samples normalized by the parameters of the target domain dataset, the overestimation of susceptibility values reached the lowest point.

6.6 Summary and Conclusion

In this chapter, transfer learning-based methods were applied to landslide susceptibility assessment. Besides, the transfer learning based on the source domain normalization was firstly proposed in this field, which significantly improved the prediction accuracy for areas with limited data. Moreover, a nationwide dataset was used for transfer learning, which is also of the first trial. Chongqing City was selected as the target area in the case study to validate the proposed methods and hypotheses. A comparison between the DL model and TL models was conducted. The following conclusions can be drawn.

Deep learning can generate reasonable outputs when there are sufficient data points. However, for data-limited cases, it failed to capture the right features and tended to overestimate the susceptibility values, which caused the "very high" susceptibility areas that dominated the generated landslide susceptibility maps. In contrast, the maps generated by transfer learning-based models were much more reasonable, which successfully presented the spatial distribution of historical landslides and areas with potential landslides.

Conventional transfer learning can effectively solve the potential difficulties of landslide susceptibility assessment caused by limited data resources. From the comparison between the DL and TL models, the prediction accuracy of the study area is improved by 41.1%, and the improvement of the loss reached 8.8% in the case of 1% of target data used for training. For the cases of 5% and 10%, the AUC improvement achieved 20.7% and 1.7%, respectively, and the loss reduction reached 24.6% and 15%, respectively. Meanwhile, it can also generate better performance in data-rich cases.

In the data-limited cases, transfer learning based on source domain normalization can conduct even better performance. In this study, the improvements of prediction accuracy achieved 51%, 24.4%, and 2.15% in the cases where 1%, 5%, and 10% of the target data used for training, respectively. The reduction of loss in each cases was 16.1%, 27.6%, and 15.4%, respectively. However, for the cases where there are enough data points, this proposal will have negative effects on landslide susceptibility assessment compared with DL model. When 95% of the target data was used for training, it resulted in a 3.2% reduction of the AUC value and a 6.2% increase of the loss. So for data-limited cases, transfer learning is suggested to be combined with source domain normalization, but for the cases where there is sufficient data, conventional transfer learning-based methods can be applied.

References

Brabb E (1984) Innovative approaches to landslide hazard mapping. In: Proceedings of the 4th international symposium landslides

Corominas J, van Westen C, Frattini P, Cascini L, Malet JP, Fotopoulou S, Catani F, van den Eeckhaut M, Mavrouli O, Agliardi F, Pitilakis K, Winter MG, Pastor M, Ferlisi S, Tofani V, Hervás J,

Smith JT (2013) Recommendations for the quantitative analysis of landslide risk. Bull Eng Geol Env. https://doi.org/10.1007/s10064-013-0538-8

Duan M, Wang G, Niu C (2018) Method of small sample size image recognition based on convolution neural network. Comp Eng Design 39(1):224–229

Fang M, Yin J, Zhu X (2013) Transfer learning across networks for collective classification. In: Proceedings of the 2013 IEEE 13th International conference on data mining, pp 161–170

Fengmei W, Jianpei Z, Yan C, Jing Y (2014) FSFP: transfer learning from long texts to the short. Appl Math Inform Sci 8(4):2033–2040. https://doi.org/10.12785/amis/080462

Gao Y, Mosalam KM (2018) Deep transfer learning for image-based structural damage recognition. Comp Aided Civ Infrastruct Eng 33(9):748–768. https://doi.org/10.1111/mice.12363

Goyes-Peñafiel P, Hernandez-Rojas A (2021) Landslide susceptibility index based on the integration of logistic regression and weights of evidence: a case study in Popayan, Colombia. Eng Geol 280:105958. https://doi.org/10.1016/j.enggeo.2020.105958

Guzzetti F, Reichenbach P, Ardizzone F, Cardinali M, Galli M (2006) Estimating the quality of landslide susceptibility models. Geomorphology 81(1–2):166–184. https://doi.org/10.1016/j.geomorph.2006.04.007

Huang F, Yin K, Huang J, Gui L, Wang P (2017) Landslide susceptibility mapping based on self-organizing-map network and extreme learning machine. Eng Geol 223:11–22. https://doi.org/10.1016/j.enggeo.2017.04.013

Ioffe S, Szegedy C (2015) Batch normalization: accelerating deep network training by reducing internal covariate shift. In: Proceedings of the 32nd international conference on machine learning, in proceedings of machine learning research, vol 37, pp 448–456. Available from https://proceedings.mlr.press/v37/ioffe15.html

Ji J, Cui H, Zhang T, Song J, Gao Y (2022) A GIS-based tool for probabilistic physical modelling and prediction of landslides: GIS-FORM landslide susceptibility analysis in seismic areas. Landslides. https://doi.org/10.1007/s10346-022-01885-9

Li Z, Zhang F, Gu W, Dong M (2020) The Niushou landslide in Nanjing City, Jiangsu Province of China: a slow-moving landslide triggered by rainfall. Landslides 17(11):2603–2617. https://doi.org/10.1007/s10346-020-01441-3

Mathew J, Jha VK, Rawat GS (2008) Landslide susceptibility zonation mapping and its validation in part of Garhwal Lesser Himalaya, India, using binary logistic regression analysis and receiver operating characteristic curve method. Landslides 6(1):17–26. https://doi.org/10.1007/s10346-008-0138-z

Nhu V-H, Shirzadi A, Shahabi H, Singh SK, Al-Ansari N, Clague JJ, Jaafari A, Chen W, Miraki S, Dou J, Luu C, Górski K, Thai Pham B, Nguyen HD, Ahmad BB (2020) Shallow landslide susceptibility mapping: a comparison between logistic model tree, logistic regression, naïve Bayes tree, artificial neural network, and support vector machine algorithms. Int J Environ Res Public Health 17(8):2749. https://doi.org/10.3390/ijerph17082749

Nian TK, Wu H, Li DY, Zhao W, Takara K, Zheng DF (2020) Experimental investigation on the formation process of landslide dams and a criterion of river blockage. Landslides 17(11):2547–2562. https://doi.org/10.1007/s10346-020-01494-4

Pathak Y, Shukla P, Tiwari A, Stalin S, Singh S, Shukla P (2022) Deep transfer learning based classification model for COVID-19 disease. IRBM 43(2):87–92. https://doi.org/10.1016/j.irbm.2020.05.003

Pires De Lima R, Marfurt K (2019) Convolutional neural network for remote-sensing scene classification: transfer learning analysis. Remote Sens 12(1):86. https://doi.org/10.3390/rs12010086

Reichenbach P, Rossi M, Malamud BD, Mihir M, Guzzetti F (2018) A review of statistically-based landslide susceptibility models. Earth Sci Rev 180:60–91. https://doi.org/10.1016/j.earscirev.2018.03.001

Rice L, Wong E, Kolter Z (2020) Overfitting in adversarially robust deep learning. In: Proceedings of the 37th international conference on machine learning research, vol 119, pp 8093–8104. Available from https://proceedings.mlr.press/v119/rice20a.html

Shao S, McAleer S, Yan R, Baldi P (2019) Highly accurate machine fault diagnosis using deep transfer learning. IEEE Trans Industr Inf 15(4):2446–2455. https://doi.org/10.1109/tii.2018.2864759

Shen SSP, Tafolla N, Smith TM, Arkin PA (2014) Multivariate regression reconstruction and its sampling error for the Quasi-global annual precipitation from 1900 to 2011. J Atmos Sci 71(9):3250–3268. https://doi.org/10.1175/jas-d-13-0301.1

Sheng L, Wang W, Zhu W (2016) China statistical yearbook 2016. China Statistics Press, Beijing (in Chinese)

Stott P (2016) How climate change affects extreme weather events. Science 352(6293):1517–1518. https://doi.org/10.1126/science.aaf7271

Sun D, Xu J, Wen H, Wang D (2021a) Assessment of landslide susceptibility mapping based on Bayesian hyperparameter optimization: a comparison between logistic regression and Random Forest. Eng Geol 281:105972. https://doi.org/10.1016/j.enggeo.2020.105972

Tonini M, Pecoraro G, Romailler K, Calvello M (2020) Spatio-temporal cluster analysis of recent Italian landslides. Georisk: Assess. Manage. Risk Eng Syst Geohazards 1–19

Van Westen C, Van Asch TW, Soeters R (2006) Landslide hazard and risk zonation—why is it still so difficult? Bull Eng Geol Environ 65:167–184

Vrbancic G, Podgorelec V (2020) Transfer learning with adaptive fine-tuning. IEEE Access 8:196197–196211. https://doi.org/10.1109/access.2020.3034343

Wang H, Zhang L, Luo H, He J, Cheung R (2021a) AI-powered landslide susceptibility assessment in Hong Kong. Eng Geol 288:106103. https://doi.org/10.1016/j.enggeo.2021.106103

Wang H, Zhang L, Wang L, He J, Luo H (2021b) An automated snow mapper powered by machine learning. Remote Sens 13(23):4826

Wei F, Zhang J, Chu Y, Yang J (2014) FSFP: Transfer learning from long texts to the short. Appl Math Inform Sci 8(4):2033–2044

Weiss K, Khoshgoftaar TM, Wang D (2016) A survey of transfer learning. J Big Data 3(1). https://doi.org/10.1186/s40537-016-0043-6

Wen B, Wang S, Wang E, Zhang J (2004) Characteristics of rapid giant landslides in China. Landslides 1(4):247–261. https://doi.org/10.1007/s10346-004-0022-4

Yan L, Xu W, Wang H, Wang R, Meng Q, Yu J, Xie WC (2019) Drainage controls on the Donglingxing landslide (China) induced by rainfall and fluctuation in reservoir water levels. Landslides 16(8):1583–1593. https://doi.org/10.1007/s10346-019-01202-x

Yang B, Yin K, Lacasse S, Liu Z (2019) Time series analysis and long short-term memory neural network to predict landslide displacement. Landslides 16(4):677–694. https://doi.org/10.1007/s10346-018-01127-x

Yang CM, Chang JM, Hung CY, Lu CH, Chao WA, Kang KH (2022) Life span of a landslide dam on mountain valley caught on seismic signals and its possible early warnings. Landslides 19(3):637–646. https://doi.org/10.1007/s10346-021-01818-y

Zhang W, Li H, Han L, Chen L, Wang L (2022) Slope stability prediction using ensemble learning techniques: a case study in Yunyang County, Chongqing, China. J Rock Mech Geotech Eng. https://doi.org/10.1016/j.jrmge.2021.12.011

Zhao Y, Wang R, Jiang Y, Liu H, Wei Z (2019) GIS-based logistic regression for rainfall-induced landslide susceptibility mapping under different grid sizes in Yueqing, Southeastern China. Eng Geol 259:105147. https://doi.org/10.1016/j.enggeo.2019.105147

Zhou X, Wen H, Zhang Y, Xu J, Zhang W (2021) Landslide susceptibility mapping using hybrid random forest with GeoDetector and RFE for factor optimization. Geosci Front 12(5):101211. https://doi.org/10.1016/j.gsf.2021.101211

Chapter 7
Displacement Prediction of Jiuxianping Landslide Using GRU Networks

7.1 Introduction

China is one of the countries with the most serious landslides in the world (e.g., Wang et al. 2004; Wang et al. 2019a, b; Xing et al. 2019; Zhang et al. 2020; Gu et al. 2021). Landslides have significant destructive power and caused a large number of casualties and property losses in mountainous areas (Xie et al. 2019). Since the first impoundment in 2003, there have been more than 5000 landslides or potential landslides induced in the TGRA, where steep slopes exist, and flood occurs frequently (Gu et al. 2017). Limited by techniques and financial resources, it is impossible to take actions to mitigate all the risks related to landslides; thus, developing a reliable system that can provide early warnings and accurate predictions is a more feasible solution (e.g., Du et al. 2013; Zhou et al. 2016, 2018a, b; Huang et al. 2017). Landslide displacement is always considered a key factor to reveal and evaluate slope stability. In the TGRA, displacement of most landslides is recorded by GPS monitor point. Therefore, how to make full use of historical information to predict landslide displacement with acceptable errors is a challenging task.

In the past few decades, many researchers have contributed to developing models for understanding and predicting landslide displacement (e.g., Jibson 2007; Teza et al. 2007; Matsuura et al. 2008; Huang et al. 2017; Zhou et al. 2018a, b; Xu and Niu 2018; Yang et al. 2019). Generally, these methods can be divided into two main categories: physical and phenomenological models (Calvello et al. 2008). The former one tries to explore the relationship between the physical properties of geo-materials and slope movement. For instance, Wang and Lin (2011) conducted a shaking table model slope test to explore displacement inside the slope, and on the surface under the effect of earthquake, results determined from this model show good consistency with observed displacements. A physically-based model shows its superiority in exploring the relationship between deformation and failure mechanism. However, these models require detailed information of engineering properties, which are always involved with uncertainty and complexity (Wang et al. 2019c). Consequently, predicted displacements of

Z. Wengang et al., *Application of Machine Learning in Slope Stability Assessment*,
https://doi.org/10.1007/978-981-99-2756-2_7

complex and large landslides via physically-based models often differ greatly from monitored values.

Phenomenological models predict landslide displacement by establishing regression models, in which observed displacements and other influential factors were considered comprehensively. Because these models do not require physical and engineering properties of concerned landslides, they have a wider applicability range with higher feasibility than physical models. Generally, there are three categories of phenomenological models applied for predicting landslide displacement: empirical models, statistical models, and intelligent models. Since Saito (1965) proposed an empirical equation for predicting landslide displacement, empirical models have been considered a powerful tool for this task. For example, Saygili and Rathje (2008) developed empirical predictive models for earthquake-induced sliding displacement displacements, combined with peak ground acceleration and peak ground velocity. In terms of statistical models, Li et al. (2012) applied a linear combination model with optimal weight on landslide displacement prediction, in which displacement datasets from two landslides were integrated, enhancing model interpretability and applicability. However, equations developed by these two methods suffer from an inherent limitation, which refer both to the uncertainty of computation and the uncertainty of independent variables (Shihabudheen et al. 2017).

In contrast, intelligent models supported by machine learning algorithms gain increasing popularity because they excel at exploring complex relationships hidden in data. Recently, the application of intelligent models on landslide displacement prediction has been explored by many studies. Lian et al. (2014) proposed an ensemble of extreme learning machine model to predict the displacement of Baishuihe landslide in the TGRA. The gray relational analysis and time-series extraction were also utilized to develop this model, and predicted results show close consistency with observed displacements. Du et al. (2013) utilized a back-propagation neural network model for the displacement prediction of two typical colluvial landslides in the TGRA, which was developed based on relationship analysis and feature selection. In order to reduce computational complexity, Shihabudheen et al. (2017) used an extreme learning adaptive neuro-fuzzy inference system, combined with empirical mode decomposition technique, to predict the landslide displacement affected by rainfall and reservoir water level. Han et al. (2021) focused on the sharp change of landslide displacement and proposed a hybrid machine learning prediction model based on support vector machine techniques, including support vector classification and support vector regression, optimized by particle swarm optimization (PSO). In their work, the lag fluctuation of groundwater level was considered to achieve better results.

Most of the existing intelligent models only consider static relationships between selected predictors and landslide displacement. To better understand and present the dynamic process of landslide evolution, a main sub-category of deep learning method, RNN, is considered a reliable and powerful tool (e.g., Xu and Niu 2018; Xie et al. 2019; Yang et al. 2019). Due to specially designed recurrent connections, this method can connect historical information and then improve model performance. However, in practice, RNN suffers from the problem of gradient vanishing or exploding. A recently developed method, GRU, is proposed to solve these problems (Chung et al. 2014). Compared with conventional RNN, GRU has two gates inside: update gate deciding if calculated data should be updated and reset gate judging how important past information is. Note that landslide displacement contains time-series information, which is appropriate for developing GRU models. However, the studies on the prediction of landslide displacements using GRU are rarely reported.

This study tries to extend the GRU model to landslide displacement prediction. For illustration, the proposed approach is applied to the displacement prediction of Jiuxianping landslide, and a comparative study is conducted to compare the prediction performance of GRU, ANN, random forest regression (RFR), and multivariate adaptive regression splines (MARS). The remainder of this part starts with the introduction of four machine learning techniques, followed by a brief description of the Jiuxianping landslide. Finally, the performance of the above four machine learning models in the displacement prediction is systematically investigated.

7.2 Machine Learning Techniques

7.2.1 Multivariate Adaptive Regression Splines

MARS is a data-driven statistical method, capable of establishing the mapping between the input variables and the corresponding output variables without any assumptions for their potential interactive relationships (e.g., Friedman 1991; Zhang and Goh 2013, 2016; Goh et al. 2018; Wang et al. 2020a; Deng et al. 2021). MARS is a nonlinear and non-parametric regression method based on the piecewise strategy, which means that the MARS model is composed of a series of piecewise linear splines with different gradients. According to Friedman (1991), the end points of a spline are called knots. A piecewise linear spline between any two adjacent knots is named as a basic function (BF).

The basic functions are obtained by a stepwise search strategy, and an adaptive regression algorithm is adopted to determine the locations for all knots. Let y denotes the predictions and $X = (X_1, X_2, ..., X_n)$ be an input vector containing n variables. Then, the predicted y from MARS can be expressed by

$$y = f(X) + e = f(X_1, X_2, \dots, X_n) + e \tag{7.1}$$

where$f(X)$ is the MARS model, composed of a series of BFs, and e denotes the model error. The format of BFs could be piecewise linear or piecewise cubic functions. In this study, only the piecewise linear functions in the form of $\max(0, x - t)$ are adopted, where t represents a knot. Then, a BF can be a spline function or the product of two or more spline functions. For simplicity, at most two orders are adopted for the MARS model in this study. Thus, the MARS model with N BFs would be

$$f(X) = \beta_0 \sum_{i=1}^{N} \beta_i \cdot \mathrm{BF}_i(X) \tag{7.2}$$

where the constant coefficient β can be estimated by the least-squares method.

The process of establishing an ideal MARS model contains two sub-periods, namely a forward selection and a backward deletion procedure, respectively. The former forward selection is mainly intended to train the optimal MARS model with the smallest model error e. It deserves noting that the number of BFs should avoid reaching the predefined maximum value, otherwise it may result in an overfitting model. To this end, the backward deletion procedure is conducted to remove the redundant variables with the lowest contribution to the model through the generalized cross-validation technique (Friedman 1991).

7.2.2 Random Forest Regression

RFR is a tree-based ensemble learning algorithm with the base evaluator of several decision trees (Merghadi et al. 2020; Zhang et al. 2021a, b). The core principle of RFR is to combine the predicted results of many decision trees to provide an ensembled result. To be specific, the final forecast results can be obtained by averaging the predictions from all decision trees, which may be more reliable and convincing in many applications compared with a single decision tree. As to RFR used in this study, the base evaluator of several regression trees (RTs) can be further specified. For every branch of an RT, the mean of the samples from the leaf nodes will be calculated. The RTs will continue growing until the mean square error between each sample reaches the minimum or no more features are available. In order to obtain an ideal RFR model, two key parameters should be optimized, i.e., the number of regression trees (n_estimators) and maximum depth of the regression tree (max_depth).

To be intuitive, Fig. 7.1 demonstrates the flowchart of random forest regression. First, N samples are randomly divided into N sets of sub-samples by the bootstrap sampling technique. Then, each single regression tree is established based on every set of sub-samples. And the final predicted results are obtained by averaging the predictions of these N RTs.

7.2.3 Artificial Neural Network

ANN is a kind of conduction structure composed of a series of neurons and connectors, which is similar to the information transmission process of human neurons (e.g., Shadloo et al. 2020; Zheng et al. 2020). ANN has become one of the most useful tools in engineering applications due to its excellent performance in fitting nonlinear multi-variable problems. ANN is also named as multi-layer perception for

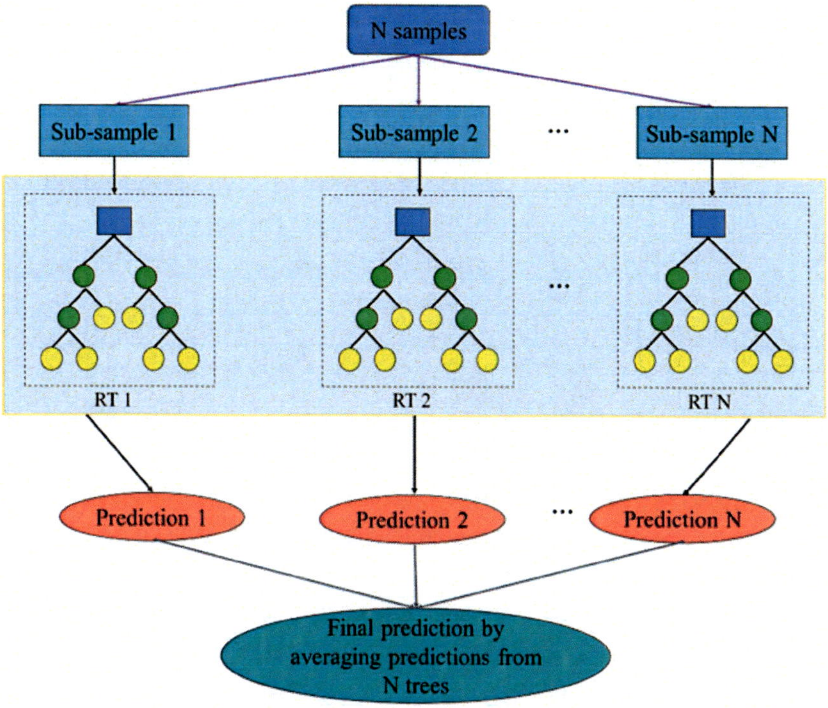

Fig. 7.1 Illustration of random forest regression with N regression trees

its structure composed of input layer, hidden layer, and output layer. Each neural in the input layer denotes an input variate. The hidden layer could contain multiple layers, depending on the complexity of the data and the accuracy of the model.

Figure 7.2 portrays the typical architecture of ANN with just one hidden layer (Zhang and Goh 2016). It is obvious that any two adjacent layers of ANN are fully connected, which means anyone neuron in the previous layer is connected to all neurons in the next layer. As shown in the figure, W_{1j} ($j = 1, 2, \ldots, q$) and W_{2k} ($k = 1, 2, \ldots, n$) are the weight associated with the connection between the entry signal and the neural. b_j ($j = 1, 2, \ldots, q$) and b_k ($k = 1, 2, \ldots, n$) are the corresponding bias.

Therefore, the output of the neural in the hidden layer and output layer can be calculated by Eqs. (7.3) and (7.4), respectively.

$$Z_j = f\left(\sum_{i=1}^{m} (W_{1i} X_i) + b_j\right), \quad j = 1, 2, \ldots, q \tag{7.3}$$

$$y_k = f\left(\sum_{j=1}^{q} (W_{2j} Z_j) + b_k\right), \quad k = 1, 2, \ldots, n \tag{7.4}$$

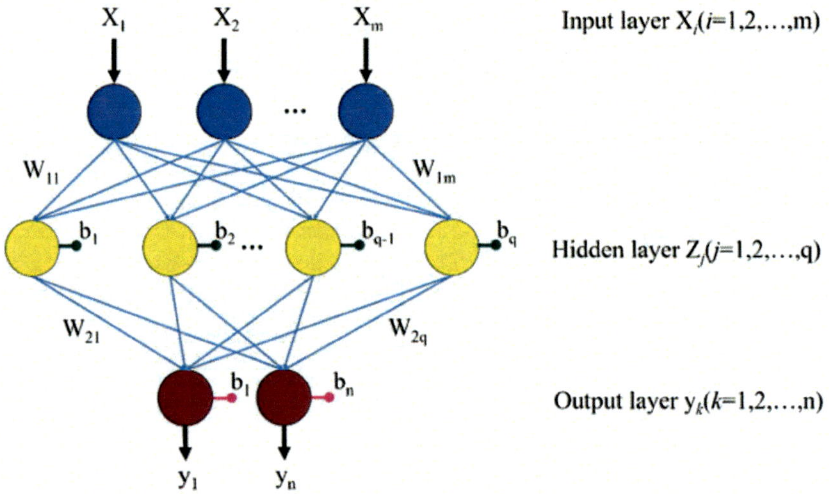

Fig. 7.2 Architecture of artificial neural network with one hidden layer

where $f(\cdot)$ is an activation function. The common activation functions contain sigmoid and tanh function, which can be expressed in Eq. (7.5) and Eq. (7.6), respectively.

$$\text{sigmoid}(x) = \frac{1}{1 + e^{-x}} \tag{7.5}$$

$$\tanh(x) = \frac{e^x - e^{-x}}{e^x + e^{-x}} \tag{7.6}$$

To obtain an optimal ANN model, the key process can be divided into two parts. The main intend in the first period is to pass the input information from several hidden layers to the output layer and randomly assign weights to each neuron. After the forward propagation is completed, there will inevitably be some deviations between the predicted result and the real value. Therefore, it is essential to optimize the current ANN model. In the second stage, the predicted output is approximated to the expected output by updating the weight of each neuron of each layer with the back-propagation algorithm. In the process of back-propagation, the gradient descent method is generally adopted to minimize the model error by calculating the derivative of the error.

7.2.4 Gated Recurrent Unit

GRU is a variant of RNN. Compared with RNN, GRU is more simplified but with enhanced performance (Cho et al. 2014; Zhao et al. 2018). Similar to RNN and long short-term memory (LSTM), GRU can also be adopted to perform the time-series predictions. The model structure of GRU is shown in Fig. 7.3, containing just a reset gate $\mathbf{r}_t$ and an update gate $\mathbf{z}_t$. The reset gate $\mathbf{r}_t$ controls how much information of new input $\mathbf{x}_t$ from the previous state will be retained to the current hidden layer vector $\widetilde{\mathbf{h}}_t$. And the update gate $\mathbf{z}_t$ controls the preservation of the previous hidden layer $\mathbf{h}_{t-1}$. GRU has fewer gates than LSTM, which contributes to its high efficiency. The transition functions of GRU are formulated as follows:

$$\mathbf{r}_t = \sigma(\mathbf{W}_{rx}\mathbf{x}_t + \mathbf{W}_{rh}\mathbf{h}_{t-1} + \mathbf{p}_r) \tag{7.7}$$

$$\mathbf{z}_t = \sigma(\mathbf{W}_{zx}\mathbf{x}_t + \mathbf{W}_{zh}\mathbf{h}_{t-1} + \mathbf{p}_z) \tag{7.8}$$

$$\widetilde{\mathbf{h}}_t = \tanh(\mathbf{W}_{hx}\mathbf{x}_t + \mathbf{W}_{hh}(\mathbf{r}_t \odot \mathbf{h}_{t-1}) + \mathbf{p}_h) \tag{7.9}$$

$$\mathbf{h}_t = (1 - \mathbf{z}_t) \odot \mathbf{h}_{t-1} + \mathbf{z}_t \odot \widetilde{\mathbf{h}}_t \tag{7.10}$$

$$\sigma(t) = \frac{1}{1 + e^{-t}} \tag{7.11}$$

$$\tanh(t) = (e^t - e^{-t})/(e^t + e^{-t}) \tag{7.12}$$

where $\mathbf{W}_{rx}$, $\mathbf{W}_{rh}$, and $\mathbf{P}_r$ are the weight matrices and bias terms of the reset gate $\mathbf{r}_t$. Similarly, $\mathbf{W}_{zx}$, $\mathbf{W}_{zh}$, and $\mathbf{P}_z$ are the weight matrices and bias terms of the update gate $\mathbf{z}_t$. $\mathbf{p}_h$ is the corresponding bias of the current hidden layer vector $\widetilde{\mathbf{h}}_t$.

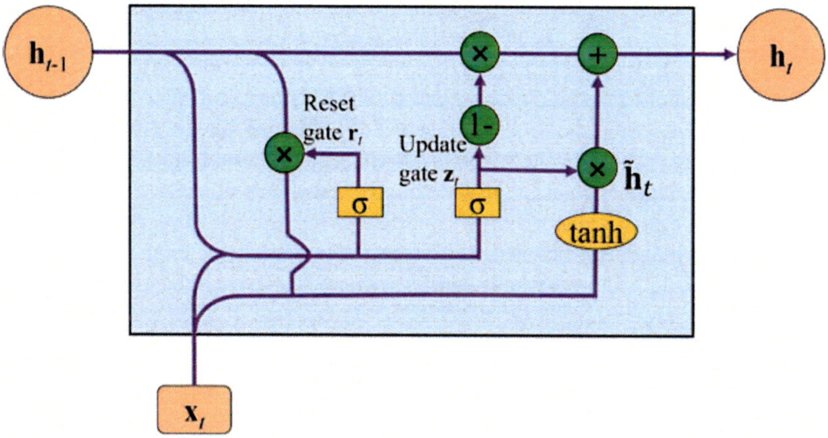

Fig. 7.3 Structure of the GRU model

7.3 Case Study: Jiuxianping Landslide

7.3.1 Geological Conditions

The Jiuxianping landslide (108° 46′ 57.35″ E, 30° 56′ 27.63″ N) is located in Fuxing Community, Qinglong Street, Yunyang County, on the left bank of the Yangtze River. The landslide area is fan-shaped, and the overall boundary conditions are clear. The volume of the landslide is about 68 million m^3, which is a super-large bedding rock landslide. Due to human engineering construction, the front edge of the landslide is the topography of a stepped platform. The middle-rear part of the landslide is an accumulated gentle slope, and it is broken terrain in the lateral direction caused by the gullies and the previous sliding. Based on the existing drilling results, the sliding mass is mainly silty clay with crushed stone (Q_4), and the sliding bed is the medium-thick gray sandstone and siltstone (J_{3s}).

The leading edge of the Jiuxianping landslide extends below the 145 m water level of the Yangtze River, and its slope stability is significantly affected by the reservoir water level. After the water level of the Three Gorges Reservoir was raised to 175 m, about 1/10 of the landslide was below the submergence line. When the reservoir water level drops from 175 to 145 m, the hydrostatic pressure of the leading edge decreases, and the hydrodynamic pressure of the landslide increases. In turn, the stability of the slope will be significantly reduced.

7.3.2 Deformation Characteristics Analysis

According to the on-site investigations, the landslide has the characteristics of slip-bending failure and has undergone three deformation stages, including slight bending stage, intense bending and bulging stage, and penetration of the sliding zone. Based on the landslide's topography, geological conditions, and deformation characteristics of the landslide, the landslide has been monitored since October 2004. There are currently 11 monitoring points. According to the monitoring data, the deformation of the leading edge of the landslide is greater than that of the trailing edge.

The data recorded by YY0210 has more obvious step characteristics, which reflects that it can better help explore the potential relationship between landslide displacement and external triggering factors. Therefore, the typical section II–II′ and the related monitoring point YY0210 were selected for specific analysis, as shown in Fig. 7.4. In addition, Fig. 7.5 shows the cumulative displacement, rainfall, and reservoir level of the monitoring point YY0210 from 2005 to 2019. The summary of the deformation characteristics related to the inducing factors is as follows:

(i) Reservoir operation is closely related to the cumulative displacement of landslides. When the reservoir water level rises or remains stable, the displacement remains unchanged. However, when the reservoir water level drops, the

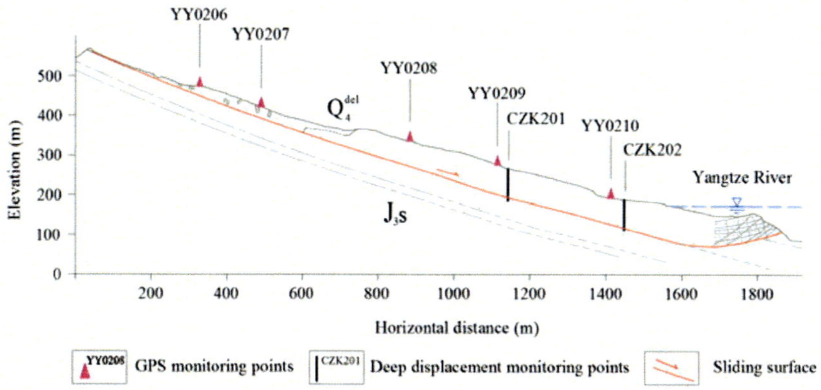

Fig. 7.4 Schematic diagram of II–II′ monitoring profile

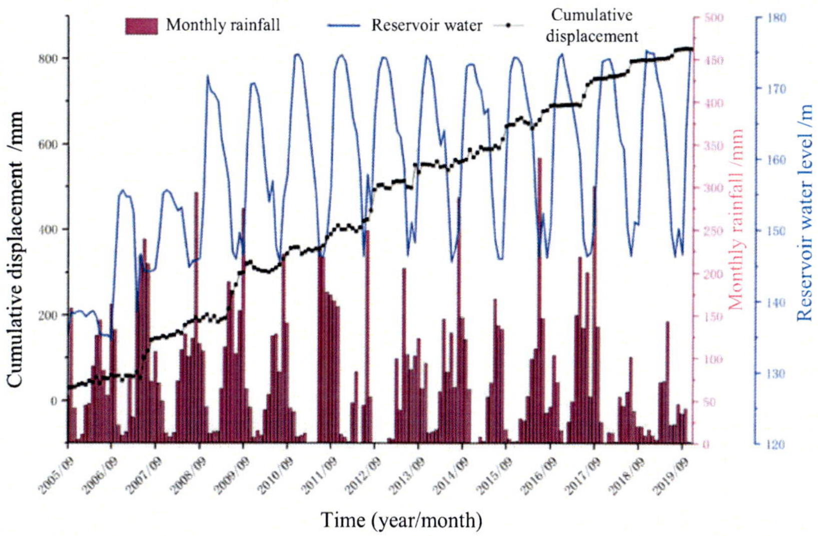

Fig. 7.5 Relationships of cumulative displacement, rainfall, and reservoir water level of monitoring point YY0210 of Jiuxianping landslide

displacement increases sharply, and the accumulated deformation increases. From 2007 to 2018, the reservoir water level changed periodically every year, and the cumulative total displacement of landslides during this period also increased accordingly. However, the deformation did not immediately respond to the drop in the water level, and the displacement often starts to increase after the reservoir water level drops for one to two months.

(ii) Rainfall also affects the stability of the landslide, although the accumulated displacement may be weakly related to this factor. In the rainy season, the cumulative displacement will increase rapidly. The reservoir water level begins to rise in late August each year; the displacement of the landslide may continue to increase and rebound due to high-intensity rainfall. This phenomenon indicates that rainfall plays a certain role in inducing landslide deformation.

(iii) Under the coupling effect of rainfall and the drop of reservoir water level, the landslide deforms sharply, and the monitored displacement has obvious stepped characteristics, indicating the influence of periodic factors. Specifically, during the rainy season from May to August, rainfall reaches its peak. At this time, the reservoir water level dropped rapidly, the deformation of the landslide suddenly increased, and the accumulated displacement curve showed a steep slope. It can be inferred that the superimposed rainfall conditions will further accelerate the displacement of the landslide during the drop of the reservoir water level.

7.3.3 Decomposition of the Cumulative Displacement

Landslide deformation is a complex nonlinear dynamic process and is influenced by internal geological conditions and several external triggering factors (e.g., rainfall infiltration and water level fluctuation) (e.g., Huang et al. 2017; Zhou et al. 2018b; Yang et al. 2019). Generally, the cumulative displacement time series can be decomposed into two components:

$$S(t) = \phi(t) + \eta(t) \tag{7.13}$$

where t is the time; $S(t)$ is the cumulative displacement; $\phi(t)$ represents the trend displacement which is influenced by the internal geological conditions on a long time and usually increases monotonically with the time; $\eta(t)$ denotes the periodic displacement that is mainly controlled by external triggering factors.

Among several displacement decomposition methods, the moving average method is adopted in this study to extract the trend displacement by virtue of its capacity to eliminate the influence of step in the cumulative displacement (e.g., Yang et al. 2019). Based on the original cumulative displacement time series $S(t) = \{s_1, s_2, \ldots, s_t\}$, the trend term $\phi(t)$ can be calculated by

$$\phi(t) = (s_t + s_{t-1} + \cdots + s_{t-n+1})/k, \quad t = k, k+1, k+2, \ldots, n \tag{7.14}$$

where s_t represents the cumulative displacement at the time t; n denotes the total number of measured points; k is the moving average cycle, and the value of 12 is used in this study.

As mentioned, the monitoring data ranging from September 2005 to December 2019 of the point YY0210 on the Jiuxianping landslide is selected in this study for the establishment and validation of machine learning models, giving rise to a database

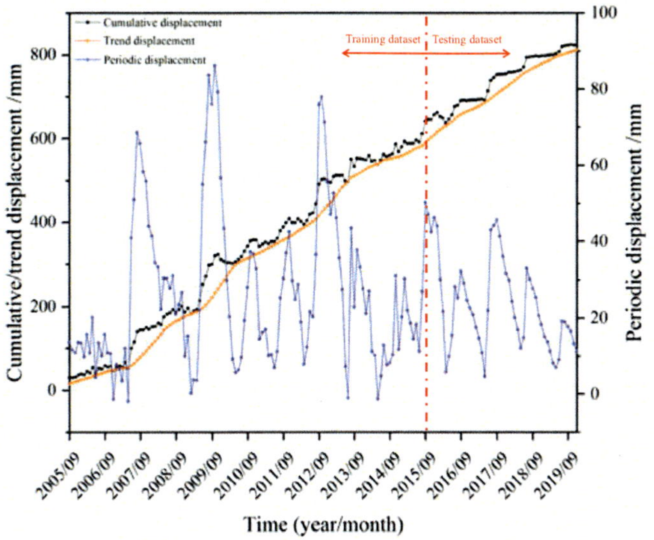

Fig. 7.6 Cumulative displacement decomposition diagram of monitoring point YY0210

consisting of 172 data points. Generally, the dataset can be divided into training dataset and testing dataset according to the commonly used ratio of 7:3. As shown in Fig. 7.6, the monitoring data from September 2005 to August 2015 is regarded as training dataset, and the remaining data from September 2015 to December 2019 is used as testing dataset.

Additionally, it is necessary to normalize the original data contained in the training dataset and testing dataset into a desired range because the machine learning models are sensitive to the ranges of data (e.g., Huang et al. 2017; Yang et al. 2019). The original data can be transformed by

$$x^{*}=\frac{x-x_{\min}}{x_{\max}-x_{\min}} \tag{7.15}$$

where x is the original value; $x_{\min}$ and $x_{\max}$ are the minimum and maximum value, respectively.

7.3.4 *Performance Measures*

Several statistical indexes can be used to evaluate the predictive performance of the established machine learning models (e.g., Zheng et al. 2019; Wang et al. 2020b; Zhang et al. 2021a, b; Li et al. 2021; Zhou et al. 2021). In this study, four indexes are adopted, namely coefficient of determination (R^2), rooted mean squared error

(RMSE), mean absolute percentage error (MAPE), and bias factor (b). Accordingly, their mathematical expressions are given below (e.g., Ching and Phoon 2014; Zhang et al. 2021a):

$$R^2 = 1 - \frac{\sum_{i=1}^{N}\left(y_i - \hat{y}_i\right)^2}{\sum_{i=1}^{N}\left(y_i - \overline{y}\right)^2} \tag{7.16}$$

$$\text{RMSE} = \sqrt{\frac{1}{N}\sum_{i=1}^{N}\left(y_i - \hat{y}_i\right)^2} \tag{7.17}$$

$$\text{MAPE} = \frac{100\%}{N}\sum_{i=1}^{N}\left|\frac{y_i - \hat{y}_i}{y_i}\right| \tag{7.18}$$

$$b = \frac{1}{N}\sum_{i=1}^{N}\frac{y_i}{\hat{y}_i} \tag{7.19}$$

where N denotes the total number of data, y_i and $\hat{y}_i$ represent the observed values and predicted values, respectively; $\overline{y}$ is the mean of the observed values. Generally, the larger the value of R^2, the better the predictive model. The closer the RMSE and MAPE values are to 0 indicating that the more accurate the predictive model is. In addition, if the bias factor b equals 1.0, which means that the predictive model is unbiased.

7.4 Results and Discussion

7.4.1 Trend Displacement Prediction

The trend displacement is influenced by internal geological conditions and usually increases monotonically with time. Based on the grid search method, a GRU model with four-layer neural network is applied to predict the trend displacement of the Jiuxianping landslide. The first three layers are the GRU layer (each layer is connected and the information is transferred between them), the fourth layer is dense layer (i.e., the layer which is fully connected with the output value), and the best step length is set to 2. Figure 7.7 compares the prediction results of the trend displacement for the training dataset and testing dataset. It is shown that the trend displacement values evaluated from the established GRU model agree well with the monitoring data for both the training dataset and testing dataset. To quantitatively evaluate the model performance, the absolute error and relative error are also calculated, as shown in Table 7.1. It can be observed that both the absolute error and relative error are marginal. In addition, Table 7.2 summarizes the results of the four evaluation indexes used in this study, further verifying the validity of the established GRU model in the trend displacement prediction of the Jiuxianping landslide.

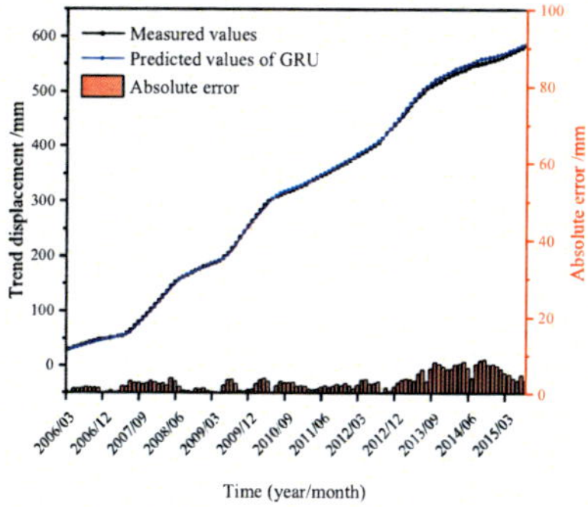

(a) Absolute error of training dataset

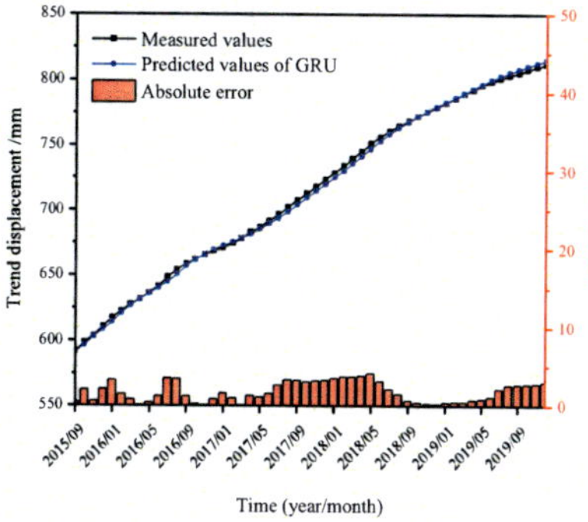

(b) Absolute error of testing dataset

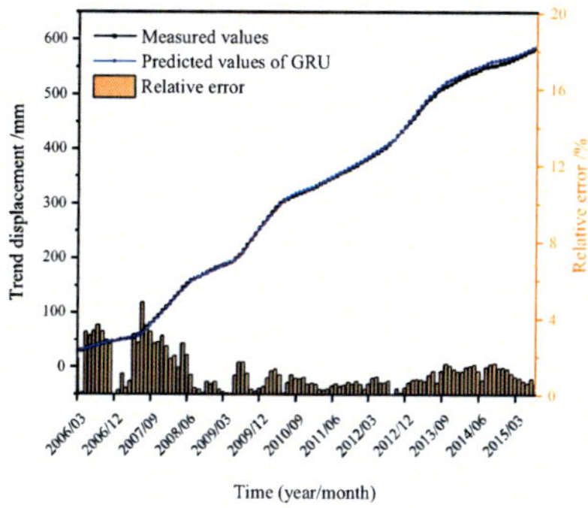

(c) Relative error of training dataset

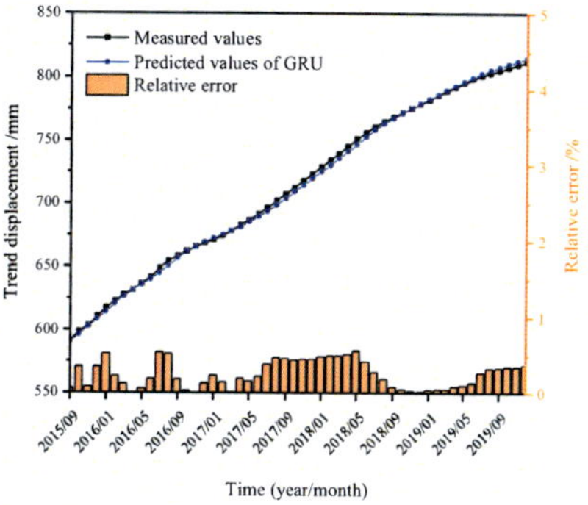

(d) Relative error of testing dataset

Fig. 7.7 Trend displacement prediction results

Table 7.1 Trend term displacement error of GRU model

Dataset	Absolute error/mm			Relative error/%		
	Minimum	Maximum	Mean	Minimum	Maximum	Mean
Training	0	8.9	2.8	0	5.7	1.2
Testing	0	4.2	1.8	0	0.6	0.2

Table 7.2 GRU model evaluation index

Evaluation index	Training dataset	Testing dataset
R^2	1.000	0.999
RMSE	3.587	2.169
MAPE	0.012	0.002
b	1.002	1.001

7.4.2 Periodic Displacement Prediction

In the periodic displacement prediction, a GRU model with four-layer neural network is constructed, where the first three layers are GRU layers, and the fourth layer is the dense layer. Moreover, the best step length is set to 6. In order to compare the prediction performance of different machine learning models, three other methods are adopted to predict the periodic displacement of Jiuxianping landslide, namely ANN, MARS, and RFR. Compared with the GRU model which is a dynamic model, all the latter three machine learning models are static models. For the ANN model, three-layer neural network structure is selected, and the number of units in each layer is 200. For the MARS model, the highest order of basis function is set to 3. For the RFR model, the *N*_estimators = 29 and Max_depth = 10 are used in this study. Figures 7.8, 7.9, 7.10, and 7.11 plot the periodic displacement prediction results of the four machine learning models for both the training dataset and testing dataset. In general, it is observed that all the four machine learning models (i.e., GRU, ANN, RFR, and MARS) can portray the monitoring data in the training dataset well.

Among the four machine learning models, the ANN and RFR model perform relatively better than the GRU and MARS model in the periodic displacement prediction

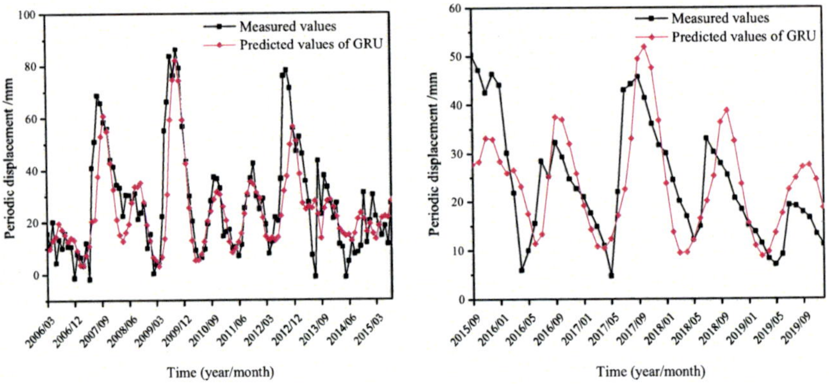

(a) Training dataset (b) Testing dataset

Fig. 7.8 Periodic displacement prediction results of GRU model

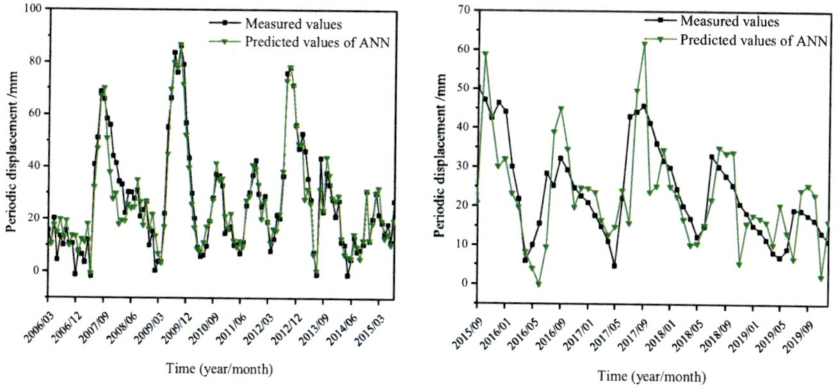

(a) Training dataset

(b) Testing dataset

Fig. 7.9 Periodic displacement prediction results of ANN model

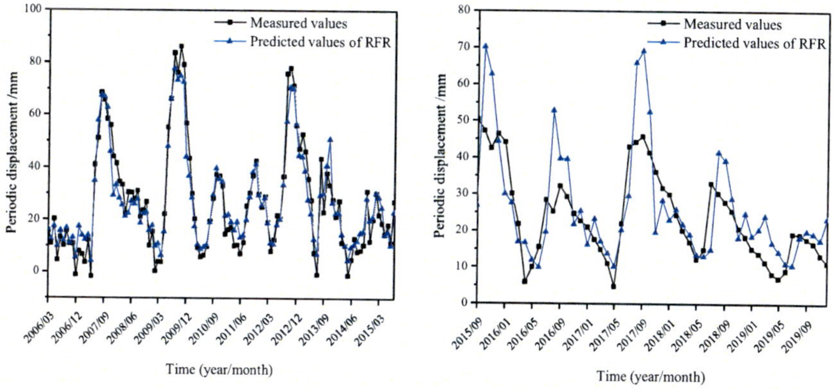

(a) Training dataset

(b) Testing dataset

Fig. 7.10 Periodic displacement prediction results of RFR model

of the training dataset. Although the GRU model can capture the changes of monitoring data in the training dataset, there exists an undesirable discrepancy between the results evaluated from the GRU model and the monitoring data during the period of 2014–2015. This discrepancy may be attributed to the poor regularity of periodic displacement between 2014 and 2015, because the predictive performance of GRU model in the current time step is closely related to the previous information. The predictive performance of the MARS model on the training dataset is not as good as the first three models (i.e., GRU, ANN, and RFR). Although the MARS model captures the peak values well, the periodic displacement values predicted from the MARS models are larger than the monitoring data for the valley points.

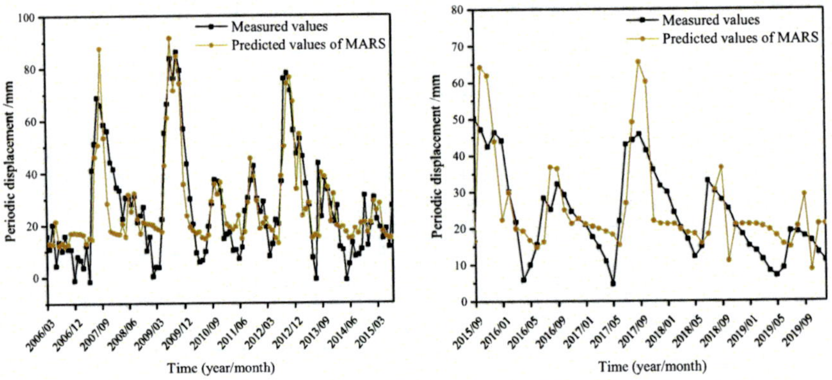

(a) Training dataset (b) Testing dataset

Fig. 7.11 Periodic displacement prediction results of MARS model

For the testing dataset, the predictive performance of the four machine learning models is generally not as good as that of the training dataset. In general, the GRU model is able to portray the variation of the periodic displacement on the testing dataset with fewer outliers. In contrast, the other three models (i.e., ANN, RFR, and MARS) are unable to predict the peak values of periodic displacement well, and the corresponding prediction errors of them are relatively larger. This indicates that the GRU model which is essentially a dynamic model making full use of the historical information can properly portray the deformation characteristics of the Jiuxianping landslide than the other three static models (i.e., ANN, RFR, and MARS).

Furthermore, Table 7.3 summarizes the results of RMSE, MAPE, and *b* calculated from the four machine learning models for both the training dataset and testing dataset. The predictive performance of the GRU model on the testing dataset is relatively satisfactory, indicating that the GRU model has stronger generalization ability and is able to capture the underlying variation tendency of the unknown data. In geotechnical engineering practice, the step-like periodic displacement may occur frequently since the periodic displacement is significantly influenced by external factors (e.g., rainfall infiltration and water level fluctuation). In such a case, it is advisable to apply the dynamic model (e.g., the GRU model) for predicting the periodic displacement, which performs better than the static models.

7.4.3 Cumulative Displacement Prediction

The final cumulative displacement can be readily obtained by taking the sum of the trend displacement and periodic displacement. Figure 7.12 compares the prediction results evaluated from the GRU model and monitoring data for the training dataset and testing dataset. It is shown that the predicted cumulative displacement generally

Table 7.3 Evaluation index results of the four prediction models

Model	Training dataset			Testing dataset		
	RMSE	MAPE	*b*	RMSE	MAPE	*b*
MARS	10.900	1.616	0.994	10.726	0.461	1.019
RFR	5.654	0.748	0.903	10.663	0.387	0.964
ANN	5.828	0.749	0.906	9.447	0.337	1.028
GRU	12.338	1.105	1.131	9.407	0.407	1.059

agrees well with the monitoring data. Although the prediction results tend to fluctuate near the stepped points of cumulative displacement, the GRU model is able to capture the whole deformation characteristics of the Jiuxianping landslide. For the training dataset, the mean absolute error and mean relative error are 9.5 mm and 4.5%, respectively. For the testing dataset, the mean absolute error and mean relative error are 8.7 mm and 1.2%, respectively. It is also observed that the maximum absolute error occurs in July 2017 (i.e., 22.9 mm), which may be attributed to the difficulty in predicting the periodic displacement during the rainy season. This indicates that the periodic displacement affects the prediction accuracy of cumulative displacement directly, especially for landslides with stepped deformation characteristics.

Table 7.4 summarizes the prediction error of cumulative displacement for the four machine learning models. For the ANN model, the mean absolute error and mean relative error of the training dataset are 6.2 mm and 3.2%, respectively. For the testing dataset, the mean absolute error and mean relative error of the training dataset are 8.1 mm and 1.1%, respectively. Figures 7.13, 7.14, and 7.15 plot the prediction results of cumulative displacement evaluated from the other three machine learning models (i.e., ANN, RFR, and MARS) and monitoring data for the training dataset and testing dataset. The prediction accuracy of the RFR is similar to the ANN, and both of them perform better than the MARS model.

In general, it shows that the GRU model can provide more satisfactory prediction results than the remaining three machine learning models (i.e., ANN, RFR, and MARS), which suggests that the dynamic model performs better than the static models in the displacement prediction of Jiuxianping landslide. The main reason is that the GRU model uses a specific length of the input sequence (including the previous information with a certain history) to predict the target value of the current time step. Different from the other three machine learning models, the GRU model can make full use of historical information based on the novel design of reset gate and update gate. When new information is available, the model can combine the stored rules with the previous information to achieve better performance. By repeating this process, the model can delete the invalid rules and keep the useful rules to update the learning relationship. This novel learning mechanism makes the GRU model portray the dynamic process of landslide deformation well.

Nonetheless, many parameters need to be determined in the GRU model construction, which is a cumbersome and time-consuming task. In addition, the prediction performance of the GRU model is influenced by the size of the data. The limited data

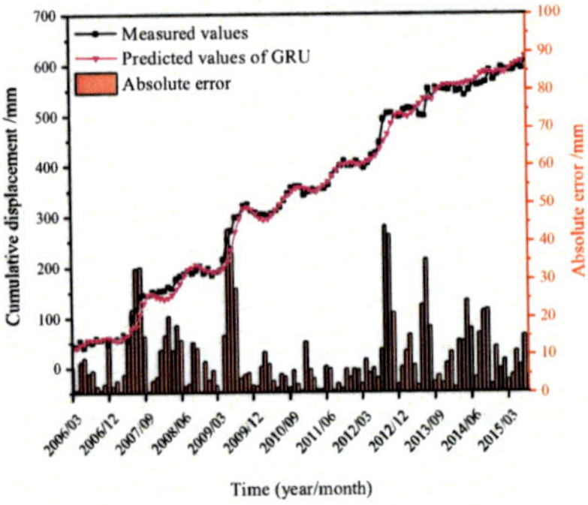

(a) Absolute error of training dataset

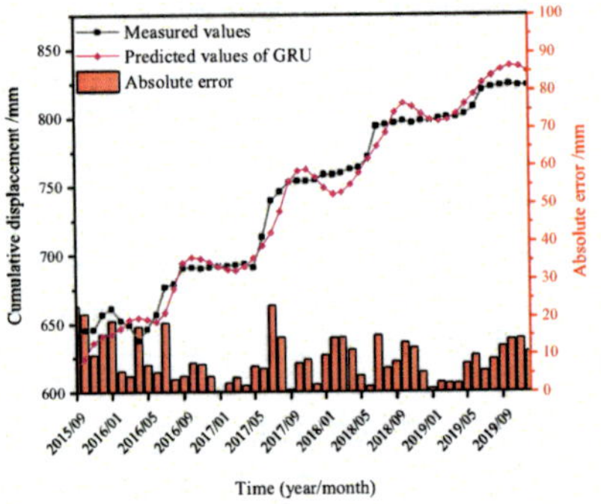

(b) Absolute error of testing dataset

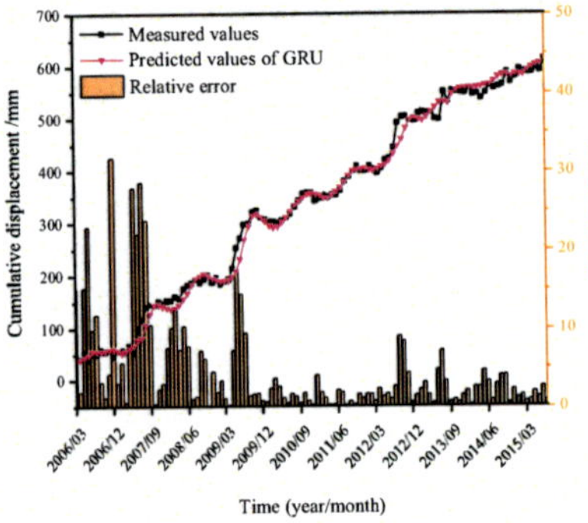

(c) Relative error of training dataset

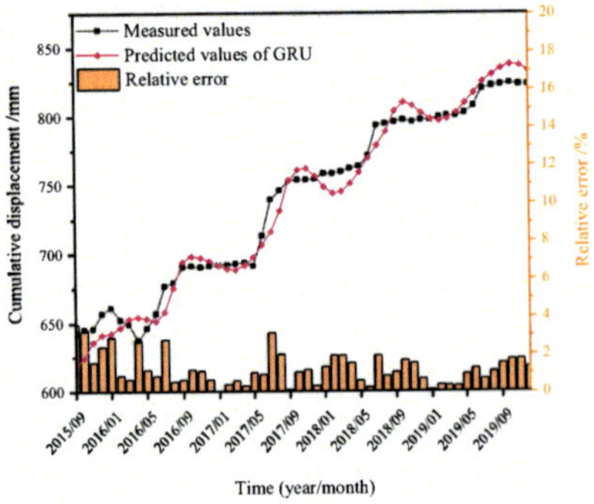

(d) Relative error of testing dataset

Fig. 7.12 Cumulative displacement prediction results of GRU model

Table 7.4 Cumulative displacement prediction error of the four machine learning models

Model	Dataset	Absolute error/mm			Relative error/%		
		Minimum	Maximum	Mean	Minimum	Maximum	Mean
GRU	Training	0.1	44.1	9.5	0.1	31.6	4.5
	Testing	0.1	22.9	8.7	0	3.6	1.2
ANN	Training	0	21.7	6.2	0	33.2	3.2
	Testing	0.6	28.7	8.1	0.1	4.5	1.1
MARS	Training	0.1	30.3	10.1	0	40	5.3
	Testing	0.2	34	8.8	0	5.3	1.2
RFR	Training	0	25.3	6.0	0	17	2.7
	Testing	1.6	24.2	8.6	0.2	3.7	1.2

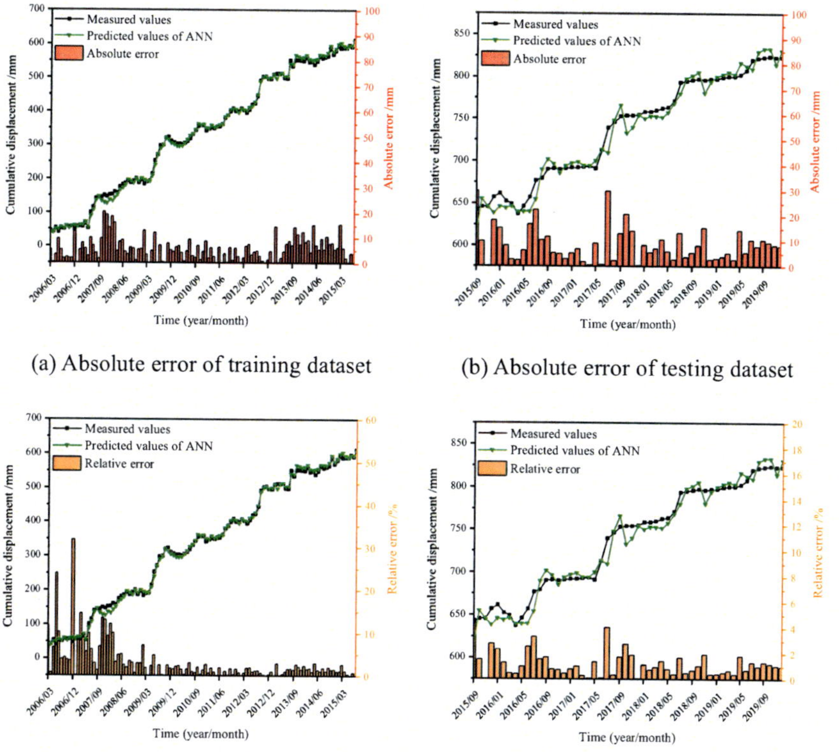

(a) Absolute error of training dataset
(b) Absolute error of testing dataset
(c) Relative error of training dataset
(d) Relative error of testing dataset

Fig. 7.13 Cumulative displacement prediction results of ANN model

will restrict the training process of the deep machine learning model, thus affecting the prediction accuracy of landslide displacement. Although the interpolation method is commonly used to increase the number of data, the interpolation points may not necessarily be consistent with the real data and thus may be unable to reflect the real deformation trend in future, which eventually leads to undesirable errors in the landslide displacement prediction. Thus, it is advisable to focus more on data preparation in geotechnical and geological engineering applications.

7.5 Summary and Conclusions

This chapter developed a GRU-based method for landslide displacement prediction. The proposed approach introduces an advanced deep machine learning method called GRU to predict the landslide displacement, which provides a novel way to portray the dynamic evolution characteristics of landslide deformation. The proposed approach

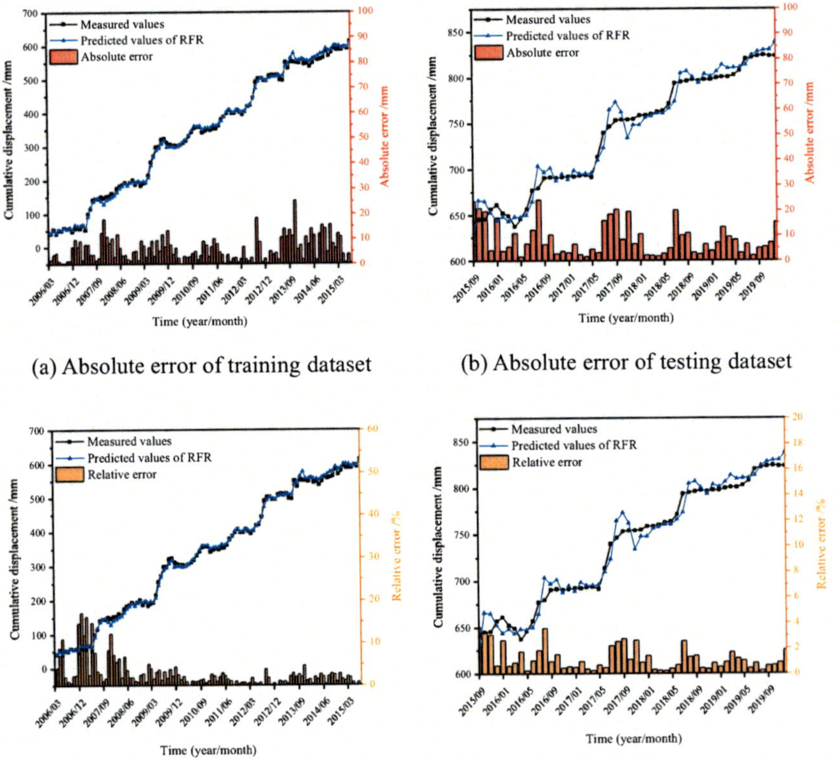

(a) Absolute error of training dataset

(b) Absolute error of testing dataset

(c) Relative error of training dataset

(d) Relative error of testing dataset

Fig. 7.14 Cumulative displacement prediction results of RFR model

was applied to the displacement prediction of the Jiuxianping landslide, and a comparative study is conducted to compare the prediction performance of GRU, ANN, RFR, and MARS. The following conclusions can be drawn from this chapter:

1. The proposed GRU-based approach performs well in the Jiuxianping landslide displacement prediction. Besides, it is shown that the GRU model can provide more satisfactory prediction results than the remaining three machine learning models (i.e., ANN, RFR, and MARS), indicating that the dynamic model performs better than the static models in the displacement prediction of Jiuxianping landslide. It provides a possibility of integrating advanced machine learning algorithms into landslide displacement prediction for rationally capturing the deformation trend in practical applications.
2. In the prediction of periodic displacement, the GRU model is able to capture the deformation trend of the testing dataset with reasonable accuracy. In contrast, the other three models (i.e., ANN, RFR, and MARS) are unable to predict the peak values of periodic displacement well, and the corresponding prediction

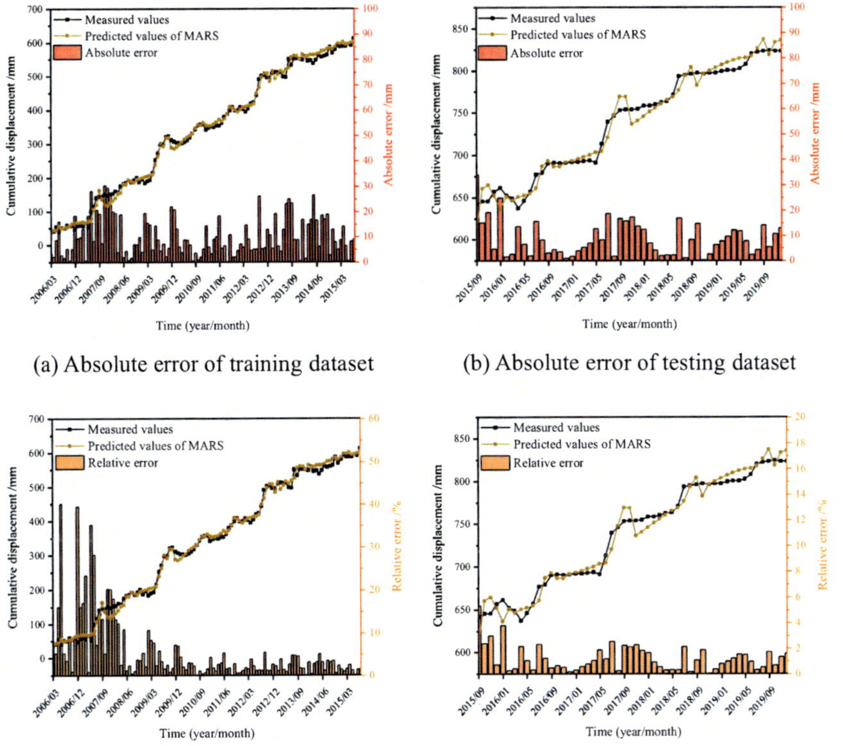

(a) Absolute error of training dataset (b) Absolute error of testing dataset

(c) Relative error of training dataset (d) Relative error of testing dataset

Fig. 7.15 Cumulative displacement prediction results of MARS model

errors of them are relatively larger. Besides, it can be observed that the ANN model achieves a better performance than the RFR and MARS. In general, it is preferable to apply the dynamic model (e.g., the GRU model) for predicting the periodic displacement of the Jiuxianping landslide, which performs better than the static models.

3. The main advantages of GRU over other static models (e.g., ANN, RFR, and MARS) are its capacity to make full use of the historical displacement information to predict the displacement at the next moment. Different from the other three machine learning models, the GRU model establishes the link between historical information and the current task based on the novel design of reset gate and update gate, allowing it to update the rules by remembering useful information and deleting invalid rules. This novel learning mechanism makes the GRU model portray the dynamic process of landslide deformation well. Nonetheless, the limited data will restrict the training process of the deep machine learning model, thus affecting the prediction accuracy of landslide displacement. Thus,

it is advisable to pay more attention to the data preparation in geotechnical and geological engineering applications.

References

Calvello M, Cascini L, Sorbino G (2008) A numerical procedure for predicting rainfall-induced movements of active landslides along pre-existing slip surfaces. Int J Numer Analyt Meth Geomech 32:327–351

Ching J, Phoon K (2014) Transformations and correlations among some clay parameters—the global database. Canadian Geotech J 51(6):663–685

Cho K, van Merriënboer B, Gulcehre C, Bougares F, Schwenk H, Bengio Y (2014) Learning phrase representations using RNN encoder-decoder for statistical machine translation

Chung J, Gulcehre C, Cho K, Bengio Y (2014) Empirical evaluation of gated recurrent neural networks on sequence modeling. arXiv preprint arXiv:1412.3555

Deng ZP, Pan M, Niu JT, Jiang SH, Qian WW (2021) Slope reliability analysis in spatially variable soils using sliced inverse regression-based multivariate adaptive regression spline. Bull Eng Geol Environ (0123456789)

Du J, Yin K, Lacasse S (2013) Displacement prediction in colluvial landslides, three Gorges reservoir, China. Landslides 10:203–218

Friedman JH (1991) Multivariate adaptive regression splines. Ann Statist 19:1–67

Goh ATC, Zhang W, Zhang Y, Xiao Y, Xiang Y (2018) Determination of earth pressure balance tunnel-related maximum surface settlement: a multivariate adaptive regression splines approach. Bull Eng Geol Environ 77:489–500

Gu D, Liu H, Gao X, Huang D, Zhang W (2021) Influence of cyclic wetting–drying on the shear strength of limestone with a soft interlayer. Rock Mech Rock Eng 0123456789

Gu DM, Huang D, Yang WD, Zhu JL, Fu GY (2017) Understanding the triggering mechanism and possible kinematic evolution of a reactivated landslide in the Three Gorges Reservoir. Landslides 14:2073–2087

Han H, Shi B, Zhang L (2021) Prediction of landslide sharp increase displacement by SVM with considering hysteresis of groundwater change. Eng Geol 280:105876

Huang F, Huang J, Jiang S, Zhou C (2017) Landslide displacement prediction based on multivariate chaotic model and extreme learning machine. Eng Geol 218:173–186

Jibson RW (2007) Regression models for estimating coseismic landslide displacement. Eng Geol 91:209–218

Li X, Kong J, Wang Z (2012) Landslide displacement prediction based on combining method with optimal weight. Nat Haz 61:635–646

Li J, Li P, Guo D, Li X, Chen Z (2021) Advanced prediction of tunnel boring machine performance based on big data. Geosci Front 12(1):331–338

Lian C, Zeng Z, Yao W, Tang H (2014) Ensemble of extreme learning machine for landslide displacement prediction based on time series analysis. Neur Comput Appl 24:99–107

Matsuura S, Asano S, Okamoto T (2008) Relationship between rain and/or meltwater, pore-water pressure and displacement of a reactivated landslide. Eng Geol 101:49–59

Merghadi A, Yunus AP, Dou J, Whiteley J, ThaiPham B, Bui DT, Avtar R, Abderrahmane B (2020) Machine learning methods for landslide susceptibility studies: a comparative overview of algorithm performance. Earth Sci Rev 207

Saito M (1965) Forecasting the time of occurrence of a slope failure. In: 6th International conference on soil mechanics and foundation engineering, pp 537–541

Saygili G, Rathje EM (2008) Empirical predictive models for earthquake-induced sliding displacements of slopes. J Geotech Geoenviron Eng 134:790–803

Shadloo MS, Rahmat A, Karimipour A, Wongwises S (2020) Estimation of pressure drop of two-phase flow in horizontal long pipes using artificial neural networks. J Energy Resour Technol Trans Asme 142

Shihabudheen KV, Pillai GN, Peethambaran B (2017) Prediction of landslide displacement with controlling factors using extreme learning adaptive neuro-fuzzy inference system (ELANFIS). Appl Soft Comput 61:892–904

Teza G, Galgaro A, Zaltron N, Genevois R (2007) Terrestrial laser scanner to detect landslide displacement fields: a new approach. Int J Remote Sens 28:3425–3446

Wang FW, Zhang YM, Huo ZT, Matsumoto T, Huang BL (2004) The July 14, 2003 Qianjiangping landslide, three gorges reservoir, China. Landslides 1:157–162

Wang KL, Lin ML (2011) Initiation and displacement of landslide induced by earthquake—a study of shaking table model slope test. Eng Geol 122:106–114

Wang LQ, Yin YP, Huang BL, Zhang ZH, Wei YJ (2019) Formation and characteristics of Guang'an village landslide in Wuxi, Chongqing, China. Landslides 16(1):127–138

Wang LQ, Yin YP, Zhang ZH, Huang BL, Wei YJ, Zhao P, Hu MJ (2019) Stability analysis of the Xinlu village landslide (Chongqing, China) and the influence of rainfall. Landslides 16(10):1993–2004

Wang L, Wu C, Gu X, Liu H, Mei G, Zhang W (2020) Probabilistic stability analysis of earth dam slope under transient seepage using multivariate adaptive regression splines. Bull Eng Geol Environ 79(6):2763–2775

Wang L, Wu C, Tang L, Zhang W, Lacasse S, Liu H, Gao L (2020) Efficient reliability analysis of earth dam slope stability using extreme gradient boosting method. Acta Geotech 15(11):3135–3150

Wang Y, Tang H, Wen T, Ma J (2019) A hybrid intelligent approach for constructing landslide displacement prediction intervals. Appl Soft Comput 81:105506

Xie P, Zhou A, Chai B (2019) The application of long short-term memory (LSTM) method on displacement prediction of multifactor-induced landslides. IEEE Access 7:54305–54311

Xing Y, Yue J, Chen C, Cong K, Zhu S, Bian Y (2019) Dynamic displacement forecasting of dashuitian landslide in china using variational mode decomposition and stack long short-term memory network. Appl Sci 9:2951

Xu S, Niu R (2018) Displacement prediction of Baijiabao landslide based on empirical mode decomposition and long short-term memory neural network in three Gorges area, China. Comp Geosci 111:87–96

Yang B, Yin K, Lacasse S, Liu Z (2019) Time series analysis and long short-term memory neural network to predict landslide displacement. Landslides 16(4):677–694

Zhang W, Tang L, Li H, Wang L, Cheng L, Zhou T, Chen X., 2020. Probabilistic stability analysis of Bazimen landslide with monitored rainfall data and water level fluctuations in Three Gorges Reservoir, China. Front Struct Civ Eng 1–15

Zhang WG, Goh ATC (2013) Multivariate adaptive regression splines for analysis of geotechnical engineering systems. Comp Geotech 48:82–95

Zhang WG, Goh ATC (2016) Multivariate adaptive regression splines and neural network models for prediction of pile drivability. Geosci Front 7:45–52

Zhang WG, Wu CZ, Zhong HY, Li YQ, Wang L (2021) Prediction of undrained shear strength using extreme gradient boosting and random forest based on Bayesian optimization. Geosci Front 12:469–477

Zhang W, Li H, Li Y, Liu H, Chen Y, Ding X (2021) Application of deep learning algorithms in geotechnical engineering: a short critical review. Artif Intell Rev. https://doi.org/10.1007/s10462-021-09967-1

Zhao R, Wang DZ, Yan RQ, Mao KZ, Shen F, Wang JJ (2018) machine health monitoring using local feature-based gated recurrent unit networks. IEEE Trans Indust Electron 65:1539–1548

Zheng G, Yang PB, Zhou HZ, Zeng CF, Yang XY, He XP, Yu XX (2019) Evaluation of the earthquake induced uplift displacement of tunnels using multivariate adaptive regression splines. Comp Geotech 113:103099

Zheng YZ, Shadloo MS, Nasiri H, Maleki A, Karimipour A, Tlili I (2020) Prediction of viscosity of biodiesel blends using various artificial model and comparison with empirical correlations. Renew Energy 153:1296–1306
Zhou C, Yin K, Cao Y, Ahmed B (2016) Application of time series analysis and PSO–SVM model in predicting the Bazimen landslide in the Three Gorges Reservoir, China. Eng Geol 204:108–120
Zhou C, Yin K, Cao Y, Ahmed B, Li Y, Catani F, Pourghasemi HR (2018) Landslide susceptibility modeling applying machine learning methods: a case study from Longju in the three Gorges Reservoir area, China. Comp Geosci 112:23–37
Zhou C, Yin KL, Cao Y, Intrieri E, Ahmed B, Catani F (2018) Displacement prediction of step-like landslide by applying a novel kernel extreme learning machine method. Landslides 15(11):2211–2225
Zhou J, Qiu Y, Armaghani DJ, Zhang W, Li C, Zhu S, Tarinejad R (2021) Predicting TBM penetration rate in hard rock condition: a comparative study among six XGB-based metaheuristic techniques. Geosci Front 12(3):101091

Chapter 8 Efficient Seismic Stability Analysis of Slopes Subjected to Water Level Changes Using Gradient Boosting Algorithms

8.1 Introduction

The embankment is one of the important infrastructures distributed around the world and has gained increasing attention in geotechnical and hydrogeological communities because its failure may induce disastrous consequences (e.g., Hicks and Li 2018; Wang et al. 2018; Gordan et al. 2021). Rational stability assessment of embankments is a prerequisite for disaster prevention and reduction, and the index of the factor of safety (FS) obtained from deterministic slope stability analysis methods (e.g., limit equilibrium method and finite element method) is frequently applied to measure the slope stability due to its conceptual simplicity. It is well recognized that embankment slope stability is significantly affected by the combined effects of several internal factors (e.g., shear strength parameters and hydraulic parameters) and external factors (e.g., earthquakes, water level fluctuations, and rainfall). Under such circumstances, slope stability prediction can offer a fast estimation of the stability status and further provide a scientific basis for decision-making in disaster mitigation (Qi and Tang, 2018).

In the past few decades, many researchers have contributed to slope stability prediction, and significant progress has been achieved (e.g., Sakellariou and Ferentinou 2005; Gordan et al. 2016; Mahdiyar et al. 2017; Koopialipoor et al. 2019; Mojtahedi et al. 2019; Zhou et al. 2019; Bui et al. 2020; Luo et al. 2021; Zeng et al. 2021). For example, Sakellariou and Ferentinou (2005) introduced neural networks to predict slope stability. The geotechnical and geometrical parameters were taken as inputs, and the FS or stability status was considered as output in their study. Gordan et al. (2016) developed a hybrid prediction model for predicting the FS of homogeneous slopes through combining the PSO and ANN. They found that the proposed PSO-ANN method performs better than the ANN model in the prediction of FS. Mahdiyar et al. (2017) employed Monte Carlo technique to predict the FS of slopes under seismic conditions based on the five important input parameters, including slope height, slope angle, cohesion, angle of internal friction, and peak ground acceleration. Results showed that the Monte Carlo-based approach is

Z. Wengang et al., *Application of Machine Learning in Slope Stability Assessment*,
https://doi.org/10.1007/978-981-99-2756-2_8

able to predict the FS appropriately. Qi and Tang (2018) compared the predictive performance of six machine learning algorithms (i.e., logistic regression, decision tree, random forest, gradient boosting machine, support vector machine, and multi-layer perceptron neural network) and concluded that integrated artificial intelligence techniques had great potential in the prediction of slope stability.

Recently, Koopialipoor et al. (2019) compared the performance of four hybrid intelligent models in the stability prediction of slopes under static and dynamic conditions. It was observed that the PSO-ANN model was superior to the remaining three hybrid intelligent models in predicting the FS of slopes. Mojtahedi et al. (2019) proposed a Monte Carlo-based probabilistic approach for forecasting the FS of slopes and found that the internal friction angle was the most influential factor among the four inputs through conducting sensitivity analysis. Zhou et al. (2019) applied a GBM approach to predict the stability status of slopes based on an updated database that records a total of 221 historical cases gathered from the literature. They found that the proposed GBM classifier can accurately capture the nonlinear relationship between slope stability status and the six influential factors. Bui et al. (2020) presented an optimized ANN model for predicting the FS of slopes by introducing the Levenberg–Marquardt backpropagation technique. Luo et al. (2021) proposed a new hybrid intelligent model to analyze the slope stability in open-pit mines by combining the PSO and cubist algorithm, and results indicated that the proposed model was able to provide satisfactory performance in the prediction of FS. Zeng et al. (2021) investigated the predictive performance of three hybrid least squares support vector machine models and found that both the gravitational search algorithm and whale optimization algorithm could improve the predictive accuracy.

It can be observed that previous research focused more on geometric parameters, shear strength parameters, and seismic coefficients. In contrast, hydraulic parameters (e.g., saturated permeability) are rarely considered in slope stability prediction. In engineering practice, embankments are usually subjected to water level changes, which may pose potentially destabilizing effects on the embankment slope stability. Generally, the hydraulic parameters play an indispensable role in the seepage analysis and slope stability analysis, and thus, it is necessary to take the hydraulic parameters into account in the slope stability prediction of embankments. Benefited from the rapid development of artificial intelligence, many machine learning algorithms have been proposed, and they are served as a promising tool for tackling geotechnical-related issues (e.g., Zheng et al. 2019; Wang et al. 2020a, b; Deng et al. 2021; Guo et al. 2021; Ray et al. 2021; Shen et al. 2021).

This study aims to develop an efficient seismic slope stability analysis approach by introducing three advanced machine learning algorithms, namely categorical boosting (CatBoost), light gradient boosting machine (LightGBM), and XGBoost. The four influential factors (i.e., cohesion, friction angle, horizontal seismic coefficient, and saturated permeability) are selected as the inputs, and the FS is regarded as the output. The remainder of this chapter starts with the introduction of CatBoost, LightGBM, and XGBoost, followed by a description of the associated implementation procedures. Then, the proposed approach is applied to the seismic stability analysis of a hypothetical embankment example subjected to water level changes. A database consisting of 600 datasets is compiled for model calibration and evaluation, where the four influential factors are selected as the inputs and the FS is regarded as the output. Finally, the performance of CatBoost, LightBoost, and XGBoost in the prediction of FS is investigated, and the relative importance of features is ranked using the SHAP method.

8.2 Methodologies

8.2.1 *Categorical Boosting*

CatBoost is a new open source library shared by the Yandex Company, which aims to handle the categorical features and prediction shift problems in machine learning (Prokhorenkova et al. 2018; Dorogush et al. 2018). Besides numerical features, categorical features are also frequently encountered in the application of machine learning, which contains a discrete set of values that are not necessarily comparable with each other. It is evident that such categorical features cannot be identified in the binary decision trees and required to be converted to numerical features through encoding techniques. As a widely used encoding technique, the one-hot encoding may cause the curse of dimensionality in tackling the high cardinality features and tends to be more efficient in handling the low cardinality features. To address this issue, CatBoost uses the target statistics as new numerical features to deal with the categorical features, which has been proved to be the most efficient method with minimum information loss (Prokhorenkova et al. 2018). It generates a random permutation of the dataset and then calculates the average label value of the training examples with the same category in the permutation. Following Prokhorenkova et al. (2018), if $\sigma = (\sigma_1, \sigma_2, \ldots, \sigma_n)$ is a permutation, the category $x_{\sigma_p,k}$ can be substituted with the average label value $\hat{x}_{\sigma_p,k}$:

$$\hat{x}_{\sigma_p,k} = \frac{\sum_{j=1}^{p-1} [x_{\sigma_j,k} = x_{\sigma_p,k}] Y_{\sigma_j} + a \cdot P}{\sum_{j=1}^{p-1} [x_{\sigma_j,k} = x_{\sigma_p,k}] + a}, \tag{8.1}$$

where P is a prior value; a is the weight of the prior; Y_{σ_j} is a label value; and $[\cdot]$ denotes the Iverson bracket, namely $[x_{\sigma_j,k} = x_{\sigma_p,k}]$ equals 1 if $x_{\sigma_j,k} = x_{\sigma_p,k}$, and otherwise, it is equal to 0.

Traditional gradient boosting decision tree algorithms generally suffer from an inevitable problem of gradient bias, which will eventually lead to prediction shift. Although the ordered boosting algorithm can avoid the prediction shift, it may be infeasible in practical applications due to the computational complexity and memory requirements in the process of training a larger number of supporting models. In such a case, CatBoost uses a modification of the ordered boosting algorithm in which the gradient boosting algorithm with decision trees is taken as base predictors. Furthermore, CatBoost also has superiority in the aspects of fast scorer and fast training on GPU. Interested readers are referred to Prokhorenkova et al. (2018) and Dorogush et al. (2018) for more details about the CatBoost.

8.2.2 *Light Gradient Boosting Machine*

LightGBM is a novel member of the histogram-based gradient boosting decision tree (GBDT) developed by Microsoft in 2017 for tackling the problems with big data and a large number of features (Ke et al. 2017). Conventional GBDT models require scanning all the data to evaluate the information gain of all the possible split points for each feature, indicating that the computational efforts may become prohibitively expensive when the data size is large and the feature dimension is high. To address this issue, LightGBM introduces two advanced techniques called Gradient-Based One-Side Sampling (GOSS) and Exclusive Feature Bundling (EFB) to reduce the number the data instances and features in a rational manner.

The gradient of data instance generally poses a significant effect on the evaluation of information gain. Compared with the data instances with larger gradients, the data instances with small gradients contribute less to the estimation of information gain. In other words, more attention should be paid to the data instances with larger gradients. Inspired by this thought, GOSS reduces the number of data instances by excluding the data instances with small gradients and simply uses the rest to calculate the information gain. Moreover, many features may be mutually exclusive in a sparse feature space, and these mutually exclusive features are unable to take nonzero values simultaneously. The basic idea of EFB is to reduce the number of features by bundling mutually exclusive features. These two novel techniques (i.e., GOSS and EFB) enable the LightGBM to achieve excellent performance in terms of computational efficiency and memory consumption. For more detailed explanations of the LightGBM, one can refer to Ke et al. (2017).

8.2.3 *Extreme Gradient Boosting*

XGBoost is a scalable end-to-end tree boosting method developed by Chen and Guestrin (2016), which has gained increasing attention in the famous Kaggle machine learning competitions due to its advantages of high efficiency and sufficient flexibility. The main idea of XGBoost is to build classification or regression trees one by one in an additive manner, and each tree learns from its predecessors and updates the residual errors in the estimated values (Zhang et al. 2021a), as shown in Fig. 8.1. Specifically, the prediction result of the gradient boosting tree model can be evaluated by integrating the values calculated from all the previously trained trees. The depth

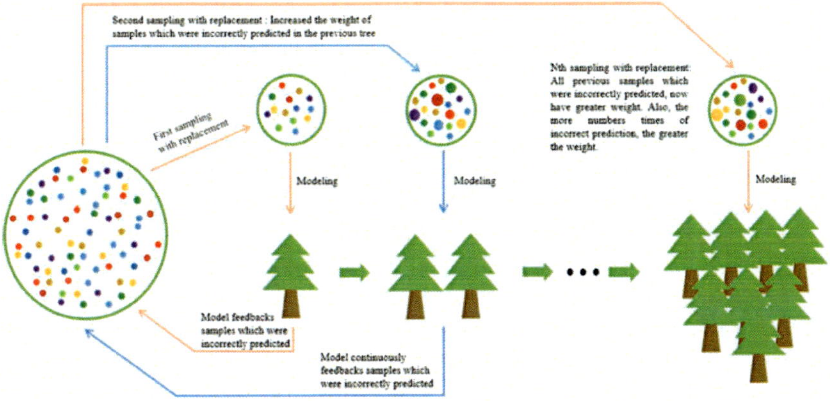

Fig. 8.1 Flowchart of the XGBoost

and the number of trees play a significant role in the XGBoost model construction, which affect the predictive accuracy directly and can be determined by optimizing the objective function. Inspired by Chen and Guestrin (2016), the objective function $\text{Obj}^{(t)}$ is expressed as:

$$\text{Obj}^{(t)} = \sum_{i=1}^{t} l(y_i, \hat{y}_i^{(t)}) + \sum_{i=1}^{t} \Omega(f_i), \tag{8.2}$$

where y_i is the actual value; $\hat{y}_i^{(t)}$ is the predicted value; $l(y_i, \hat{y}_i^{(t)})$ is the loss function describing that how well the model fits training data; and $\Omega(f_i)$ is a regularization term to penalize model complexity and avoid potential overfitting problems. For more detailed information about the XGBoost algorithm, interested readers can refer to Chen and Guestrin (2016).

8.3 Implementation Procedure

Figure 8.2 shows the implementation procedures of seismic stability analysis of embankment slopes using gradient boosting algorithms. Firstly, the database used for model calibration should be prepared, which contains the necessary information about the input parameters (e.g., shear strength parameters) and output quantity of interest (e.g., FS). Then, divide the database into the training dataset and testing dataset according to a rational ratio. Thereafter, the three variants of the gradient boosting algorithms, namely CatBoost, LightGBM, and XGBoost, are used to construct the machine learning models, where the associated hyperparameters can be determined by optimization techniques (e.g., Bayesian optimization). Finally, the

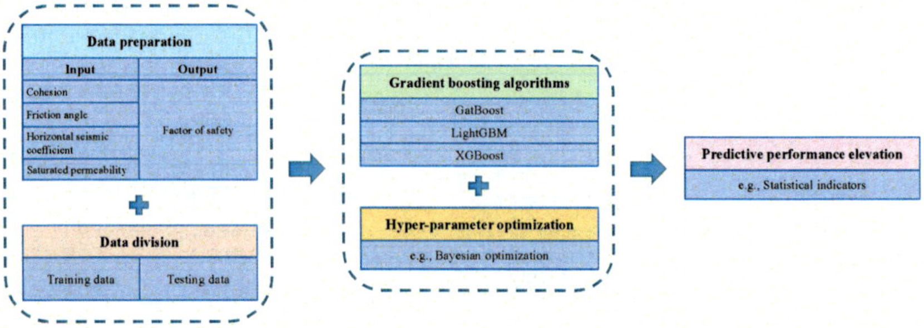

Fig. 8.2 Implementation procedure of the proposed method

predictive performance of these constructed machine learning models can be quantitatively measured using statistical indicators (e.g., the coefficient of determination R^2). For illustration, the proposed approach is applied to the seismic stability analysis of a hypothetical embankment case in the next section.

8.4 Illustrative Example

For illustration, a hypothetical embankment example with a height of 12 m and a slope of 27°is used in this study for illustration, as shown in Fig. 8.3. It is situated on a foundation of 100 m. Due to the fact that the embankments suffer from water level changes frequently, and thus a constant total head equal to the upstream water level is applied to the embankment below the water level. For the foundation, a zero flux boundary is assigned to both sides and the bottom. In this example, the 2D limit equilibrium slope stability software Slide2 (Rocscience Inc. 2018) is applied to perform seepage and slope stability analysis of the embankment example under combined effects of seismic loading and water level changes. The water level is assumed to rise uniformly from the initial water level (i.e., 17 m) to the highest water level (i.e., 19 m) after 8 days. Table 8.1 gives the mean values of the four main influential factors that govern the stability of embankment slopes, including the cohesion c, friction angle φ, horizontal seismic coefficient K_h, and saturated permeability k_s. Based on these mean values, the simplified Bishop method embedded in the Slide2 software can be applied to calculate the FS of the downstream slope. Figure 8.4a, b plots the FS values of embankment slope example at the initial state and 50 days, respectively. The stability of slopes generally requires some time to achieve a steady state after the water level reaches a designed elevation. After a series of trial calculations, the FS at 50 days is satisfied with the requirement. Thus, the FS at the 50 days is used in the following database preparation.

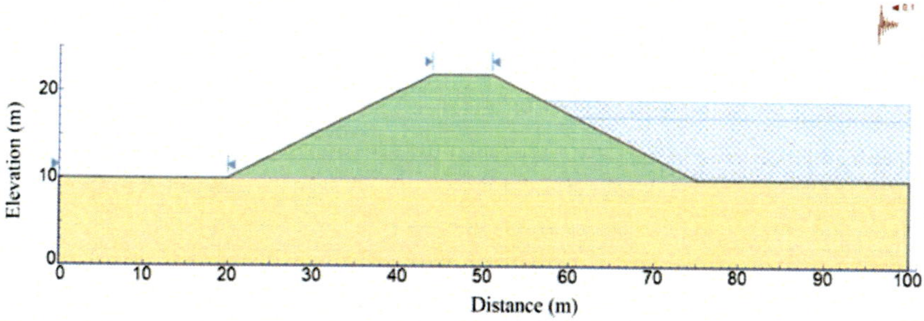

Fig. 8.3 Geometry and boundary conditions of the embankment example

Table 8.1 Statistical properties of parameters used in this example

Parameters	Mean value	COV	Distribution
Cohesion c (kPa)	8	0.25	Lognormal
Friction angle φ (∘)	28	0.15	Lognormal
Horizontal seismic coefficient K_h	0.1	0.3	Lognormal
Saturated permeability k_s (m/s)	1.0×10^{-6}	0.5	Lognormal

8.4.1 Database Preparation for Model Calibration

A database containing the four input parameters (i.e., c, φ, K_h, and k_s) and the corresponding output of FS should be prepared for calibrating the machine learning models. Inspired by previous research (e.g., Cho 2012; Li et al. 2015; Zhang et al. 2021b), the four input parameters are assumed to follow a lognormal distribution. Based on the mean values, coefficients of variation (COVs), and probability distributions given in Table 8.1, a total of 600 groups of data are generated using the Latin hypercube sampling method. Figure 8.5a–d plots the histogram of the cohesion c, friction angle φ, horizontal seismic coefficient K_h, and saturated permeability k_s, respectively. The possible ranges for c, φ, K_h, and k_s are [3.74 kPa, 17.66 kPa], [17.29°, 45.08°], [0.04, 0.25], and [1.73×10^{-7}, 5.69×10^{-6}], respectively. Each data group containing the c, φ, K_h, and k_s is used as input in the Slide2 software for calculating the FS of the embankment slope example. With the aid of Slide2 software, all the FS values corresponding to the 600 groups of data can be evaluated. As plotted in Fig. 8.5e, the FS values range from 0.747 to 1.507. These input parameters and output consequences constitute a database with a total of 600 datasets, and each dataset consists of four input parameters (i.e., c, φ, K_h, and k_s) associated with the corresponding FS value. Although the Slide2 software is used in this study to perform seismic stability analyses of the 600 groups of data, other geotechnical commercial software of interest can also be applied.

The compiled database can be divided into training dataset and testing dataset for model construction and evaluation. In this study, 400 groups of data are used as the

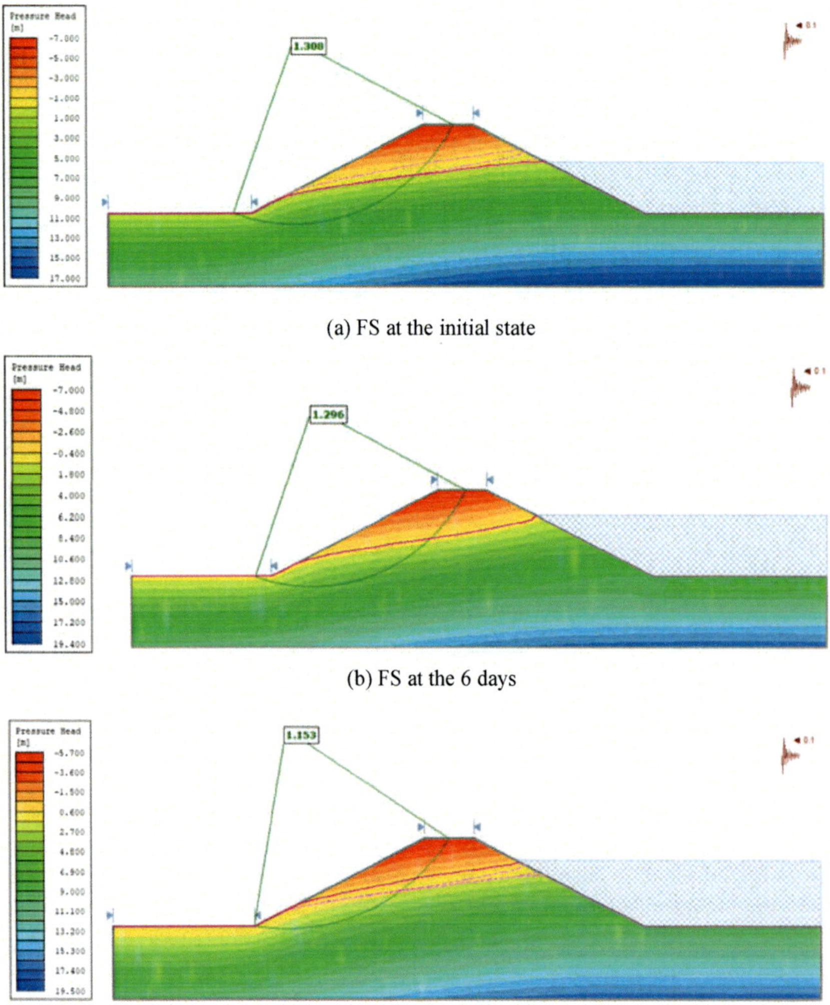

(a) FS at the initial state

(b) FS at the 6 days

(c) FS at the 50 days

Fig. 8.4 FS values of the embankment example at different times

training dataset, and 200 groups of data are regarded as the testing dataset. Then, the three gradient boosting algorithms (i.e., CatBoost, LightGBM, and XGBoost) are used to construct the machine learning models. The performance of different machine models in the prediction of FS can be evaluated using statistical indicators.

8.4.1.1 Predictive Performance of Different Models

Figure 8.6a compares the FS values obtained from the established CatBoost model and actual values calculated from the Slide2 software for all the 600 groups of data. It can be observed that the predicted FS values obtained from the established CatBoost model agree well with those calculated from the Slide2 software for both the training

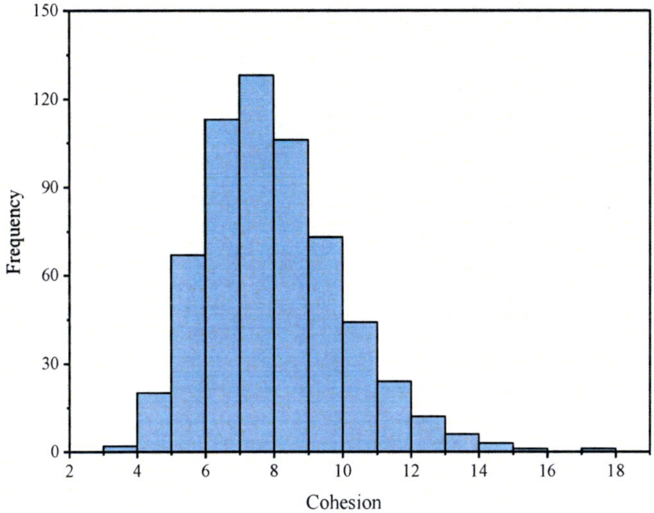

(a) c

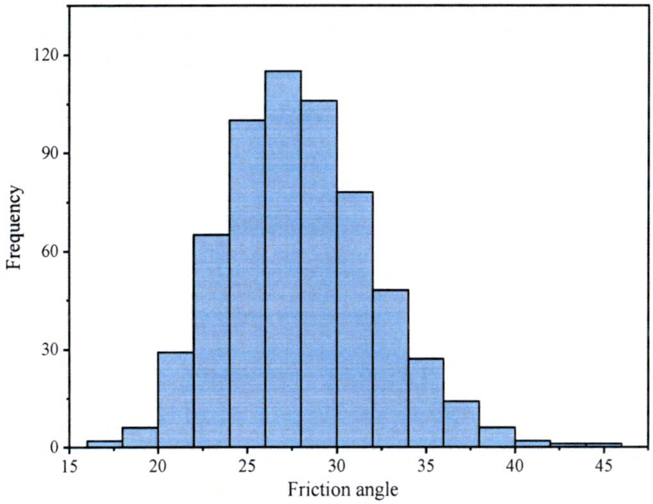

(b) φ

Fig. 8.5 Histogram of the four influential factors and factor of safety

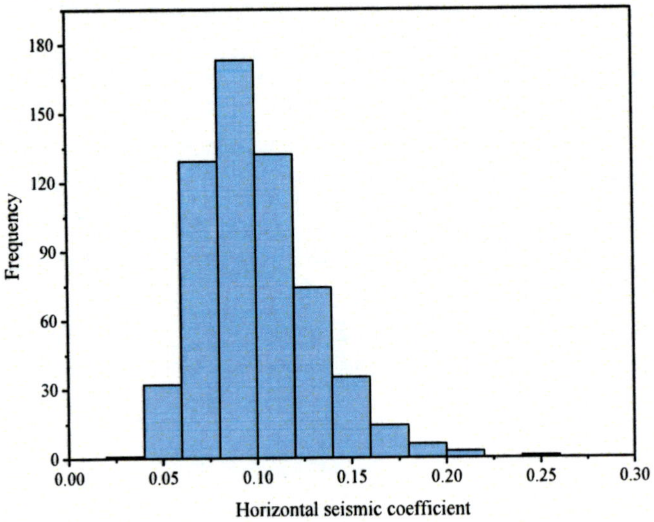

(c) K_h

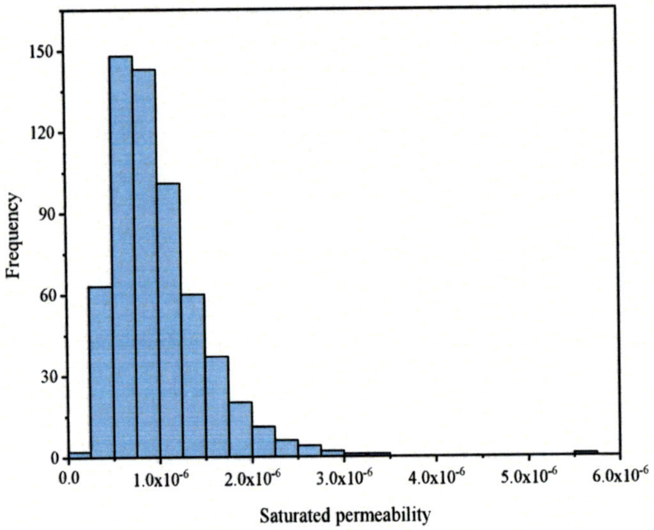

(d) k_s

Fig. 8.5 (continued)

dataset (i.e., 400 groups of data) and testing dataset (i.e., 200 groups of data). To quantitatively evaluate the model performance, the frequently used index of R^2 is used in this study. As shown in Fig. 8.6a, the R^2 values of training dataset and testing dataset are larger than 0.90, indicating that the established CatBoost model is able to

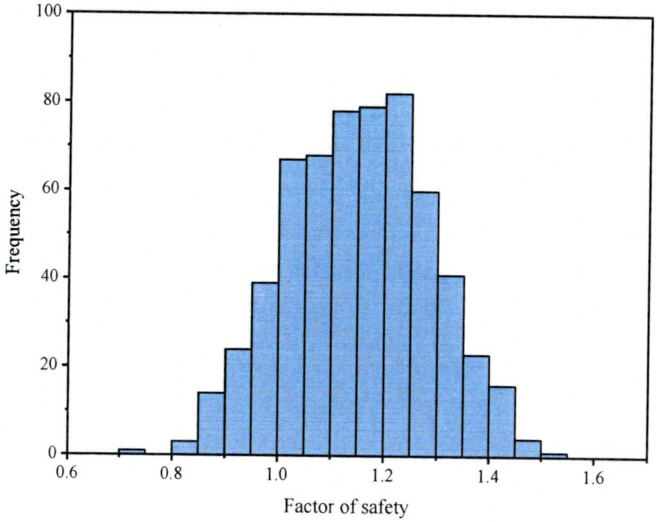

(e) FS

Fig. 8.5 (continued)

predict the FS of the embankment slope example with satisfactory accuracy. Likewise, Fig. 8.6b compares the FS values predicted from the constructed LightGBM model and actual values calculated from the Slide2 software. Both the training dataset and testing dataset can achieve a relatively high R^2 value, illustrating the excellent capability of LightGBM model in predicting the FS. Furthermore, Fig. 8.6c compares the prediction results of XGBoost model and actual values calculated from the Slide2 software. It is shown that most of the points gather around the reference line (i.e., 1:1 line), and the corresponding R^2 values of training dataset and testing dataset are also relatively high. This implies that the XGBoost model performs well in the prediction of FS. In general, it can be concluded that all the three machine learning models (i.e., CatBoost, LightGBM, and XGBoost model) are able to provide satisfactory performance in the prediction of FS for the embankment slope example, which offers a promising approach for seismic stability analysis by introducing advanced gradient boosting algorithms.

8.4.1.2 Feature Importance Analysis

To investigate the relative importance of features on the predictive performance of machine learning models, the SHAP method is used in this study due to its fast implementation for tree-based models. Inspired by the coalitional game theory, the SHAP method uses the Shapley values to quantify the contribution of each feature to the prediction (Lundberg and Lee 2017; Guo et al. 2021). Generally, the features

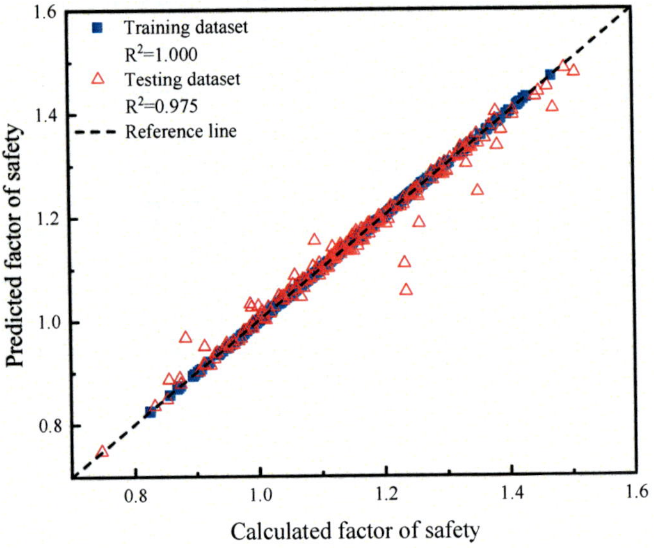

(a) CatBoost model

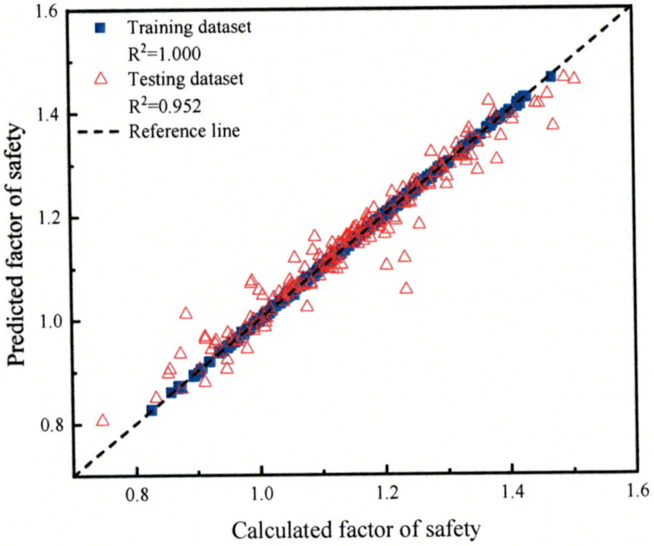

(b) LightGBM model

Fig. 8.6 Predictive performance of the three gradient boosting algorithms

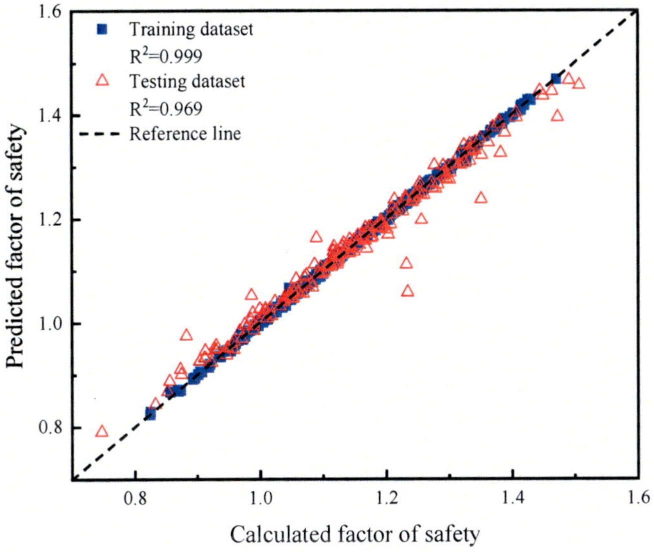

(c) XGBoost model

Fig. 8.6 (continued)

with higher positive SHAP values tend to pose a more significant influence on the final prediction. Figure 8.7 plots the SHAP values of the four features calculated from the CatBoost model. Each scattered point on the figure represents one sample, and the points with red colors indicate that the associated feature values are high. On the other hand, the blue colors imply that the feature values are low. For the friction angle φ, it can be observed that many sample points with red colors gather around the zone with positive SHAP values, indicating that the friction angle affects the FS of the embankment slopes significantly and the larger value of friction angle will enhance the embankment slope stability. In contrast, for the horizontal seismic coefficient K_h, a large number of sample points with red colors locate in the zone with negative SHAP values. This means that the horizontal seismic coefficient will weaken the embankment slope stability.

In general, it is evident that the friction angle φ has the most significant influence on the prediction of FS, followed by horizontal seismic coefficient K_h, cohesion c, and saturated permeability k_s. Among the four features, the shear strength parameters (i.e., φ and c) have positive influences on the embankment slope stability, while the increasing K_h and k_s will destabilize the embankment slope stability. Furthermore, Fig. 8.8 ranks feature importance of the four features. The arrangement of these four features from bottom to top is based on their relative importance. Similarly, it can be found that the friction angle φ has the most significant influence on the prediction of FS, followed by K_h, c, and k_s. This finding is consistent with that observed in Fig. 8.7, further validating the significance of shear strength parameters (i.e., φ and

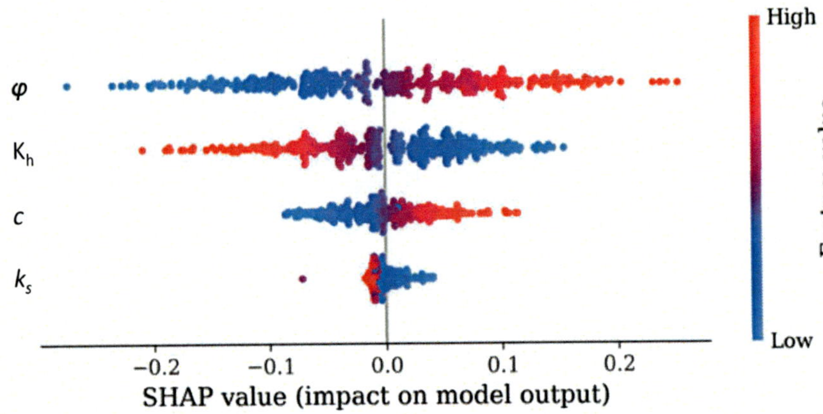

Fig. 8.7 SHAP values of features calculated from the CatBoost model

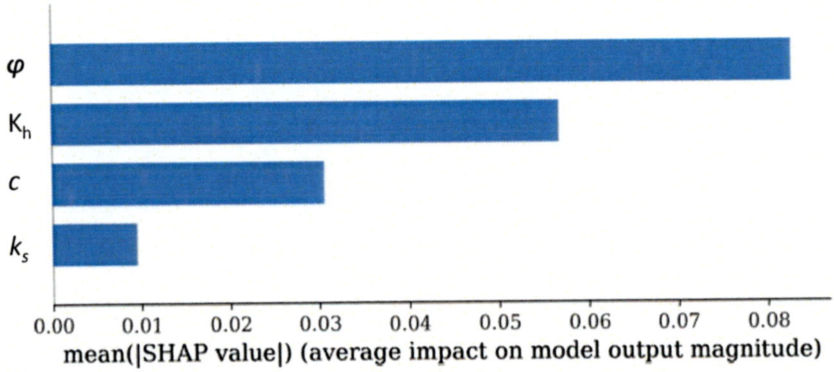

Fig. 8.8 Relative importance of features calculated from the CatBoost model

c) and seismic coefficient (i.e., K_h) in the seismic stability evaluation of embankment slopes.

8.5 Summary and Conclusions

This chapter developed a gradient boosting algorithm-based approach for seismic stability analysis of embankment slopes. Three advanced gradient boosting algorithms, namely CatBoost, LightGBM, and XGBoost, were calibrated and evaluated in this study using a well-established database that contains a total of 600 datasets. Each dataset records the four features (i.e., the cohesion, friction angle, horizontal seismic coefficient, and saturated permeability) associated with the FS. For illustration, the proposed approach was applied to the seismic stability analysis of a hypothetical embankment example subjected to water level changes. The predictive performance of CatBoost, LightBoost, and XGBoost was compared, and the relative importance of features on the prediction was also quantified by the SHAP method.

Results showed that all the R^2 values of the three gradient boosting algorithms (i.e., CatBoost, LightBoost, and XGBoost) were larger than 0.90 for both the training dataset and testing dataset, indicating that the proposed approach is able to predict the FS of embankment slopes with satisfactory accuracy. Among the four influencing factors, the friction angle φ had the most significant influence on the prediction of FS, followed by horizontal seismic coefficient K_h, cohesion c, and saturated permeability k_s. Different from the shear strength parameters (i.e., φ and c) that had positive influences on the embankment slope stability, the increasing K_h and k_s tended to destabilize the embankment slope stability. The proposed approach making the best use of advanced gradient boosting algorithms provides an efficient tool for seismic stability analysis of embankment slopes. Besides the above four influential factors, other geometric and geotechnical parameters of interest can also be considered in the future study.

References

Bui XN, Muazu MA, Nguyen H (2020) Optimizing Levenberg–Marquardt backpropagation technique in predicting factor of safety of slopes after two-dimensional OptumG2 analysis. Eng Comp 36(3):941–952. https://doi.org/10.1007/s00366-019-00741-0

Chen T, Guestrin C (2016) Xgboost: a scalable tree boosting system, In: Proceedings of the 22nd acm sigkdd international conference on knowledge discovery and data mining, 785–794

Cho SE (2012) Probabilistic analysis of seepage that considers the spatial variability of permeability for an embankment on soil foundation. Eng Geol 133–134:30–39. https://doi.org/10.1016/j.enggeo.2012.02.013

Deng ZP, Pan M, Niu JT, Jiang SH, Qian WW (2021) Slope reliability analysis in spatially variable soils using sliced inverse regression-based multivariate adaptive regression spline. Bull Eng Geol Env. https://doi.org/10.1007/s10064-021-02353-9

Dorogush AV, Ershov V, Gulin A (2018) CatBoost: gradient boosting with categorical features support. arXiv preprint, 1810.11363

Gordan B, Armaghani DJ, Hajihassani M, Monjezi M (2016) Prediction of seismic slope stability through combination of particle swarm optimization and neural network. Engineering with Computers 32(1):85–97. https://doi.org/10.1007/s00366-015-0400-7

Gordan B, Raja MA, Armaghani DJ, Adnan A (2021) Review on dynamic behaviour of earth dam and embankment during an earthquake. Geotech Geol Eng 4. https://doi.org/10.1007/s10706-021-01919-4

Guo D, Chen H, Tang L, Chen Z (2021) Assessment of rockburst risk using multivariate adaptive regression splines and deep forest model. Acta Geotech 2. https://doi.org/10.1007/s11440-021-01299-2

Hicks MA, Li Y (2018) Influence of length effect on embankment slope reliability in 3D. Int J Numer Anal Meth Geomech 42(7):891–915. https://doi.org/10.1002/nag.2766

Ke G, Meng Q, Finley T, Wang T, Chen W, Ma W, Liu TY (2017) Lightgbm: a highly efficient gradient boosting decision tree. Advances in Neural Information Processing Systems. Morgan Kaufmann Publishers, San Mateo, CA, USA, pp 3146–3154

Koopialipoor M, Jahed DJ, Hedayat A, Marto A, Gordan B (2019) Applying various hybrid intelligent systems to evaluate and predict slope stability under static and dynamic conditions. Soft Comput 23(14):5913–5929. https://doi.org/10.1007/s00500-018-3253-3

Li DQ, Jiang SH, Cao ZJ, Zhou W, Zhou CB, Zhang LM (2015) A multiple response-surface method for slope reliability analysis considering spatial variability of soil properties. Eng Geol 187:60–72. https://doi.org/10.1016/j.enggeo.2014.12.003

Lundberg SM, Lee SI (2017) A unified approach to interpreting model predictions. In: Proceedings of the 31st Conference on Neural Information Processing Systems, 4768–4777

Luo Z, Bui XN, Nguyen H, Moayedi H (2021) A novel artificial intelligence technique for analyzing slope stability using PSO-CA model. Eng Comp 37(1):533–544. https://doi.org/10.1007/s00366-019-00839-5

Mahdiyar A, Hasanipanah M, Armaghani DJ, Gordan B, Abdullah A, Arab H, Majid MZA (2017) A Monte Carlo technique in safety assessment of slope under seismic condition. Eng Comp 33(4):807–817. https://doi.org/10.1007/s00366-016-0499-1

Mojtahedi SFF, Tabatabaee S, Ghoroqi M, Tehrani MS, Gordan B, Ghoroqi M (2019) A novel probabilistic simulation approach for forecasting the safety factor of slopes: a case study. Eng Comp 35(2):637–646. https://doi.org/10.1007/s00366-018-0623-5

Prokhorenkova L, Gusev G, Vorobev A, Dorogush AV, Gulin A (2018) CatBoost: unbiased boosting with categorical features. Adv Neu Info Process Syst, 6638–6648

Qi C, Tang X (2018) Slope stability prediction using integrated metaheuristic and machine learning approaches: a comparative study. Comput Ind Eng 118:112–122. https://doi.org/10.1016/j.cie.2018.02.028

Ray R, Kumar D, Samui P, Roy LB, Goh ATC, Zhang W (2021) Application of soft computing techniques for shallow foundation reliability in geotechnical engineering. Geosci Front 12(1):375–383. https://doi.org/10.1016/j.gsf.2020.05.003

Rocscience Inc. (2018) Slide2 version 2018—2D limit equilibrium slope stability analysis, Toronto, ON, Canada. www.rocscience.com

Sakellariou MG, Ferentinou MD (2005) A study of slope stability prediction using neural networks. Geotech Geol Eng 23(4):419–445. https://doi.org/10.1007/s10706-004-8680-5

Shen H, Li J, Wang S, Xie Z (2021) Prediction of load-displacement performance of grouted anchors in weathered granites using FastICA-MARS as a novel model. Geosci Front 12(1):415–423. https://doi.org/10.1016/j.gsf.2020.05.004

Wang B, Chen YL, Wu C, Peng Y, Song JJ, Liu WJ, Liu X (2018) Empirical and semi-analytical models for predicting peak outflows caused by embankment dam failures. J Hydrol 562:692–702. https://doi.org/10.1016/j.jhydrol.2018.05.049

Wang L, Wu C, Gu X, Liu H, Mei G, Zhang W (2020a) Probabilistic stability analysis of earth dam slope under transient seepage using multivariate adaptive regression splines. Bull Eng Geol Env 79(6):2763–2775. https://doi.org/10.1007/s10064-020-01730-0

Wang L, Wu C, Tang L, Zhang W, Lacasse S, Liu H, Gao L (2020b) Efficient reliability analysis of earth dam slope stability using extreme gradient boosting method. Acta Geotech 15(11):3135–3150. https://doi.org/10.1007/s11440-020-00962-4

Zeng F, Amar MN, Mohammed AS, Motahari MR, Hasanipanah M (2021) Improving the performance of LSSVM model in predicting the safety factor for circular failure slope through optimization algorithms. Eng Comp. https://doi.org/10.1007/s00366-021-01374-y

Zhang WG, Meng FS, Chen FY, Liu HL (2021a) Effects of spatial variability of weak layer and seismic randomness on rock slope stability and reliability analysis. Soil Dyn Earthq Eng 146:106735. https://doi.org/10.1016/j.soildyn.2021.106735

Zhang W, Wu C, Zhong H, Li Y, Wang L (2021b) Prediction of undrained shear strength using extreme gradient boosting and random forest based on Bayesian optimization. Geosci Front 12(1):469–477. https://doi.org/10.1016/j.gsf.2020.03.007

Zheng G, Yang PB, Zhou HZ, Zeng CF, Yang XY, He XP, Yu XX (2019) Evaluation of the earthquake induced uplift displacement of tunnels using multivariate adaptive regression splines. Comput Geotech 113:103099. https://doi.org/10.1016/j.compgeo.2019.103099

Zhou J, Li E, Yang S, Wang M, Shi X, Yao S, Mitri HS (2019) Slope stability prediction for circular mode failure using gradient boosting machine approach based on an updated database of case histories. Saf Sci 118:505–518. https://doi.org/10.1016/j.ssci.2019.05.046

Chapter 9
Efficient Reliability Analysis of Slopes in Spatially Variable Soils Using XGBoost

9.1 Introduction

The safety of earth dam has attracted increasing concern in geotechnical engineering practice because earth dam is the most common type of dam in the world and its failure may induce considerably detrimental social and economic impacts (e.g., Hicks and Li 2018; Wang et al. 2018a). Due to the fact that a large number of the existing earth dams suffer from severe degradation and these aging earth dams give rise to a higher probability of failure (e.g., Khalilzad et al. 2015), it is of great significance to evaluate the earth dam safety for designing remedial and mitigation measures. In general, traditionally deterministic seepage and slope stability analyses are performed to evaluate the safety of earth dam via exit hydraulic gradient and FS, respectively. However, there is a growing consensus that deterministic analysis approaches are unable to account for the underlying geotechnical uncertainties (e.g., Phoon and Kulhawy 1999; Wang et al. 2016). In such a case, probabilistic analysis of geotechnical structures has received increasing attention since it provides a rational means to explicitly consider the geotechnical uncertainties (e.g., Zhao et al. 2016; Ji et al. 2018; Liu et al. 2018; Huang et al. 2018; Li et al. 2015, 2018, 2019).

In the past few decades, many researchers have performed probabilistic analysis of seepage flow through earth dam (e.g., Ahmed 2009; Cho 2012; Le et al. 2012; Liu et al. 2017) since the pioneering work by Fenton and Griffiths (1996, 1997). In contrast, little attention has been paid to the probabilistic stability analysis of earth dam slope (Gui et al. 2000; Babu and Srivastava 2010). Gui et al. (2000) attempted to investigate the influences of spatially variable saturated permeability on the earth dam slope reliability without considering the spatial variability of shear strength parameters. Although Babu and Srivastava (2010) applied first-order reliability method to the reliability analysis of earth dam slope stability, the shear strength parameters are simply regarded as random variables and their spatial variabilities are ignored. It is well recognized that the spatial variability of shear strength parameters affects the slope reliability significantly (e.g., Cho 2010; Li et al. 2015; Liu et al. 2019). Thus, besides soil permeability, it is of great importance to consider the spatial variability

Z. Wengang et al., *Application of Machine Learning in Slope Stability Assessment*,
https://doi.org/10.1007/978-981-99-2756-2_9

of shear strength parameters in the reliability analysis of earth dam slope stability. The evaluation of probability of failure P_f (or, equivalently, reliability index) is a primary concern in geotechnical reliability analysis. Monte Carlo simulation (MCS) has gained popularity in evaluating the probability of failure due to its conceptual simplicity and easy to use for geotechnical engineers. However, it usually requires extensive computational efforts to obtain an estimate of P_f with desired accuracy (Li et al. 2016a). As an alternative, many methods have been proposed to facilitate the slope reliability analysis, such as subset simulation (SS) (e.g., Wang et al. 2010, 2011), response surface method (RSM) and their advanced variants (Li et al. 2016b). The basic idea of SS is to decompose the small failure probability into the product of several intermediate events with larger failure probabilities (Au and Beck 2001; Au and Wang 2014). Huang et al. (2016) found that although the SS improves the computational efficiency, the computational efforts may still be prohibitively large in some cases. For the RSM methods, Liu and Cheng (2016) revealed that they may lead to unsatisfactory results due to the prior assumption on the form of the unknown functional relationship. In such a case, machine learning may offer a viable alternative.

Machine learning is a data-driven process of training machines to achieve expected performance based on the available data or past experience, such as MARS (e.g., Zhang and Goh 2013, 2016; Liu et al. 2019). Generally, a relatively larger number of training samples are required for calibrating a MARS model with high accuracy, and Silvestrini et al. (2013) suggested that the necessary training samples should be 10–15 times greater than the number of model parameters (i.e., random variables in reliability analysis). This may considerably increase the computational efforts and weaken the advantage of MARS in geotechnical reliability analysis applications, especially when the spatial variabilities of soil properties are considered. This calls for an efficient machine learning algorithm to facilitate the calculation of P_f in the geotechnical reliability analysis.

With the rapid advancement of computer technology, an advanced machine learning algorithm called XGBoost is recently developed by Chen and Guestrin (2016). The XGBoost is a novel extension of the commonly used gradient tree boosting, which has been widely used in machine learning and data mining competitions due to its advantages of high efficiency and sufficient flexibility. Take the famous machine learning competition sponsored by Kaggle for example, most of the winning solutions listed in the Kaggle website are conditional on XGBoost (Sheridan et al. 2016). In the field of machine learning, it is well recognized that the XGBoost is currently one of the fastest and best open source boosted tree algorithms. To the best of our knowledge, although the XGBoost has many successful applications in several fields, it is rarely applied in geotechnical engineering. The integration of the XGBoost and geotechnical reliability analysis may offer an effective way to reveal the influences of spatial variable soil properties on the earth dam slope reliability from a new perspective.

This chapter aims to propose an XGBoost-based reliability analysis approach for evaluating the earth dam slope stability efficiently. The proposed approach trains a model under tree boosting framework, and subsequently evaluates the failure probability of earth dam slope by virtue of the established XGBoost model. The remainder of this part starts with the introduction of deterministic seepage and slope stability analyses for the earth dam, followed by a description of the random field characterization of spatially variable soil properties. Then, an efficient XGBoost-based reliability analysis approach is proposed, and its implementation procedures are summarized. Finally, the proposed approach is applied to the reliability analysis of Ashigong earth dam. The influences of spatially variable soil properties on the earth dam slope failure probability are explored systematically by performing parametric sensitivity analysis.

9.2 Deterministic Analysis of Earth Dam Slope Stability

9.2.1 Seepage Analysis Under Steady Seepage Condition

Accurate analysis of seepage flow through earth dam is of great importance in geotechnical engineering, which provides valuable seepage information (e.g., flow velocity, gradient, and pore water pressure) for designing earth dam reinforcement and risk mitigation measures. Compared with the frequently encountered seepage under saturated condition, seepage flow through earth dam is a complex problem involves saturated–unsaturated state. Richards (1931) pioneered unsaturated seepage analysis by extending Darcy's law to unsaturated flow. Consider steady state flow through earth dam, the corresponding governing equation of two-dimensional flow in earth dam can be expressed as (e.g., Fredlund et al. 2012):

$$\frac{\partial}{\partial x}\left(k_x \frac{\partial H}{\partial x}\right) + \frac{\partial}{\partial y}\left(k_y \frac{\partial H}{\partial y}\right) = 0 \tag{9.1}$$

where $\partial(\cdot)/\partial x$ and $\partial(\cdot)/\partial y$ denote the first-order partial derivative on x and y, respectively; k_x and k_y are the horizontal hydraulic conductivity (i.e., in x direction) and vertical hydraulic conductivity (i.e., in y direction), respectively; H represents the total head obtained from summation of the corresponding elevation head and pressure head.

Compared with the constant hydraulic conductivity k and volumetric water content θ (i.e., saturated hydraulic conductivity k_s and saturated volumetric water content θ_s) in saturated zone, they vary with the pore-water pressure u_w (or, equivalently, matric suction ψ) for unsaturated flow. This is the primary distinction between saturated seepage and unsaturated seepage (Santoso et al. 2011). The relationship between θ and ψ can be represented by soil–water characteristic curve (SWCC), and several parametric SWCC models have been developed in the past decades (e.g., Siller and Fredlund 2001; Wang et al. 2018b, 2019a). For illustration, the van Genuchten (1980)-Mualem (1976) model (abbreviated as VGM) is used in this study. The mathematical equation of VGM model is written as (e.g., van Genuchten 1980):

$$S_e = \frac{\theta - \theta_r}{\theta_s - \theta_r} = \left[\frac{1}{1 + (\psi / \alpha_{vg})^{n_{vg}}}\right]^{m_{vg}} \quad (m_{vg} = 1 - 1/n_{vg},\ n_{vg} > 1) \tag{9.2}$$

where S_e is effective degree of saturation; θ_r and θ_s are the residual volumetric water content and saturated volumetric water content, respectively; ψ is matric suction deduced from pore-air pressure u_a and pore-water pressure u_w, namely $\psi = u_a - u_w$; α_{vg}, n_{vg} and m_{vg} are fitting parameters related to the shape of VGM model. Until now, several hydraulic conductivity functions (HCFs) have been developed to describe the relationship between hydraulic conductivity k and ψ in unsaturated soils (e.g., Leong and Rahardjo 1997). According to the statistical model theory proposed by Childs and Collis-George (1950), van Genuchten (1980) derived an HCF corresponding to VGM model, which is given by:

$$k_r(\psi) = \frac{k(\psi)}{k_s} = S_e^{1/2}\left[1 - (1 - S_e^{1/m_{vg}})^{m_{vg}}\right]^2 \tag{9.3}$$

where k_r is the relative hydraulic conductivity representing the ratio between k and k_s.

As indicated by Eqs. (9.1–9.3), it is difficult to solve the governing equation analytically due to its high nonlinearity. In such a case, numerical methods have been used as an alternative for performing seepage analysis, such as finite difference method and finite element method. Subsequently, the results of pore-water pressure and volumetric water content obtained from seepage analysis are incorporated into slope stability analysis, as discussed in the next subsection.

9.2.2 Slope Stability Analysis

The factor of safety (FS) is frequently used as an index to evaluate the slope safety in geotechnical engineering practice, which represents the minimum ratio of resisting moment and overturning moment among several potential sliding surfaces. Inspired by the findings obtained from previous experimental studies, Vanapalli et al. (1996) derived an extended Mohr–Coulomb equation for predicting the shear strength of unsaturated soils, which is given by:

$$\tau = c' + (\sigma - u_a)\tan\varphi' + (u_a - u_w)S_e\tan\varphi' \tag{9.4}$$

where τ represents the unsaturated shear strength; c' and φ' are the effective cohesion and the effective friction angle, respectively; σ represents the total normal stress; $(u_a - u_w)$ is equal to the matric suction ψ. As indicated by Eq. (9.4), the pore-water pressure u_w and effective degree of saturation S_e obtained from seepage analysis are necessary inputs in the slope stability analysis for calculating the FS. In this study, the SEEP/W module and SLOPE/W module contained in commercial software Geostudio (GEO-SLOPE International, Ltd. 2012) are employed to conduct seepage and slope stability analyses of earth dam, respectively. The deterministic seepage and slope stability analyses provide a basis for the subsequent probabilistic analyses, in which the random field characterization of spatially variable soil properties is the primary concern, as discussed in the next subsection.

9.3 Random Field Modeling of Spatially Variable Soil Properties

Rational characterization of the spatially variable soil properties is of great concern in geotechnical reliability analysis. Since the pioneering work contributed by Vanmarcke (1983), the random filed theory has gained increasing attention in geotechnical reliability analysis for considering the spatial variability of soil properties. According to the random field theory, a number of random variables are firstly generated based on the prescribed statistics (e.g., means and standard deviations) and autocorrelation information (e.g., autocorrelation structure and distance) of soil properties. Then, these random variables are mapped onto the finite element mesh for specifying the soil properties of each element in the model of geotechnical structures. This process is generally referred to as random field modeling and several random field modeling techniques have been proposed, such as midpoint method, local average subdivision, and Karhunen–Loève expansion method.

The midpoint method is applied in this study due to its advantages of conceptual simplicity and easy to implement. The correlation between different random field elements can be quantified by a predefined autocorrelation function (ACF), such as single exponential, second-order Markov, cosine exponential and binary noise ACFs. For illustration, the single exponential autocorrelation function (SECF) is employed herein, the correlation coefficient $\rho_{i,j}$ between random field elements i and j can be given by (e.g., Li et al. 2015):

$$\rho_{i,j} = \exp\left[-2\left(\frac{|x_i - x_j|}{\delta_h} + \frac{|y_i - y_j|}{\delta_v}\right)\right], \quad i, j = 1, 2, \ldots, N_e \tag{9.5}$$

where x_i and x_j are the centroid coordinates of random field elements i and j in the horizontal direction, respectively; y_i and y_j denote the centroid coordinates of random field elements i and j in the vertical direction, respectively; δ_h and δ_v represent the horizontal and vertical scales of fluctuation (SOF), respectively; N_e is the number of random field elements. After rearranging these $\rho_{i,j}$, the autocorrelation matrix $\boldsymbol{C}$ can be obtained. Subsequently, the Cholesky decomposition is applied to calculate the lower triangular matrix $\boldsymbol{L}$ of $\boldsymbol{C}$. Based on a vector $\boldsymbol{\xi}$ of standard normal random samples, the correlated standard Gaussian random field samples can be generated by multiplying $\boldsymbol{L}$ and $\boldsymbol{\xi}$. Finally, the non-Gaussian random field samples $\boldsymbol{\Theta}$ are simulated by performing isoprobabilistic transformation (e.g., Li et al. 2015). These random field samples of soil properties are then incorporated into the reliability analysis of earth dam slope stability. Note that the midpoint method is only applicable for stationary random field, however, the spatially varying soil properties generally exhibit a linear or nonlinear trend in practice. In such a case, the midpoint method may be inappropriate for portraying the spatial variability of soil properties. The newly developed random field generator called BCS-KL generator provides a versatile tool for generating the random field samples directly from the limited measurement data (e.g., Wang et al. 2018c; Hu et al. 2019), which is able to simulate the non-stationary non-Gaussian random field (Montoya-Noguera et al. 2019) and random field with unknown trend function (Wang et al. 2019b).

9.4 XGBoost-Based Reliability Analysis Approach

9.4.1 Introduction of XGBoost

XGBoost is an advanced supervised algorithm proposed by Chen and Guestrin (2016) under the tree boosting framework, which has the advantages of high efficiency and sufficient flexibility. To overcome the deficiency of single tree, a tree ensemble model constituted by several trees is employed in XGBoost to develop an input–output mapping, which can be used to explore the underlying relationship between the spatially variable soil properties and the responses of interest in geotechnical reliability analysis (e.g., FS). Let $\hat{\boldsymbol{\Theta}}_i$ denotes a set of random variables obtained from the random field modeling described above, the corresponding FS value $\hat{y}_{\mathrm{FS},i}$ predicted from XGBoost can be given by:

$$\hat{y}_{FS,i} = \sum_{k=1}^{K} f_k(\widehat{\Theta}_i), \quad f_k \in \mathcal{F} \tag{9.6}$$

where $\mathcal{F}$ is an ensemble model containing a total of K trees; $f_k(\cdot)$ is the predicted value corresponding to the k-th tree. As indicated by Eq. (9.6), $\hat{y}_{\mathrm{FS},i}$ is essentially a summation of the predicted value obtained from each tree. For a given set of training data, the total number of trees contained in $\mathcal{F}$ (i.e., K) and the optimal structure of each tree can be obtained by optimizing the predefined objective function. An objective function Γ_{obj} formulated by Chen and Guestrin (2016) contains two parts, it is expressed as:

$$\Gamma_{\mathrm{obj}} = \sum_{i=1}^{n} l(y_{\mathrm{FS},i}, \hat{y}_{\mathrm{FS},i}) + \sum_{k=1}^{K} \Omega(f_k) \tag{9.7}$$

where $l(y_{\mathrm{FS},i}, \hat{y}_{\mathrm{FS},i})$ is the loss function describing the model fit with training data; n is the total number of training data; $\boldsymbol{\Omega}(f_k)$ is an additional regularization term for penalizing complicated model and avoiding the over-fitting problem. As indicated by Eqs. (9.6) and (9.7), the optimization of $\boldsymbol{\Gamma}_{\mathrm{obj}}$ is not a trivial task because it involves a tree ensemble model. In such a case, the traditional optimization methods may be infeasible. In contrast, the additive learning strategy provides an efficient way to tackle this difficulty. All the trees are built sequentially in the additive learning processes, each newly added tree learns from its former trees and updates the residuals in the prediction values. Thus, the objective function $\boldsymbol{\Gamma}^{t}_{obj}$ for the t-th iteration is written as (Chen and Guestrin 2016):

$$\begin{aligned}\Gamma^{t}_{\mathrm{obj}} &= \sum_{i=1}^{n} l(y_{\mathrm{FS},i}, \hat{y}^{t}_{\mathrm{FS},i}) + \sum_{k=1}^{t} \boldsymbol{\Omega}(f_k) \\ &= \sum_{i=1}^{n} l(y_{\mathrm{FS},i}, \hat{y}^{t-1}_{\mathrm{FS},i} + f_t(\widehat{\boldsymbol{\Theta}}_i)) + \boldsymbol{\Omega}(f_t) + C_0 \end{aligned} \tag{9.8}$$

where C_0 denotes a constant. By virtue of the second-order Taylor expansion, the $\Gamma^{t}_{\mathrm{obj}}$ can be approximately evaluated by:

$$\Gamma^{t}_{\mathrm{obj}} \approx \sum_{i=1}^{n} \begin{bmatrix} l(y_{\mathrm{FS},i}, \hat{y}^{t-1}_{\mathrm{FS},i}) + g_i f_t(\widehat{\Theta}_i) \\ +\frac{1}{2} h_i f_t^2(\widehat{\Theta}_i) \end{bmatrix} + \Omega(f_t) + C_0 \tag{9.9}$$

where $g_i = \partial l(y_{\mathrm{FS},i}, \hat{y}^{t-1}_{\mathrm{FS},i}) / \partial \hat{y}^{t-1}_{\mathrm{FS},i}$ and $h_i = \partial^2 l(y_{\mathrm{FS},i}, \hat{y}^{t-1}_{\mathrm{FS},i}) / \partial (\hat{y}^{t-1}_{\mathrm{FS},i})^2$ denote the first and second-order partial derivatives of the loss function, respectively. Because the constant C_0 has no influence on the optimization processes, it can be removed and the $\Gamma^{t}_{\mathrm{obj}}$ can be further reformulated as:

$$\Gamma^{t}_{\mathrm{obj}} = \sum_{i=1}^{n} \left[g_i f_t(\widehat{\Theta}_i) + \frac{1}{2} h_i f_t^2(\widehat{\Theta}_i) \right] + \Omega(f_t) \tag{9.10}$$

By optimizing Eq. (9.10), the t-th tree associated with the model parameters and predictions can be determined. The optimization procedures are repeated until the predefined stopping criterion is achieved, and the ultimate predictions are subsequently obtained. More detailed explanations of the XGBoost algorithm are referred to Chen and Guestrin (2016). Based on the established XGBoost model, the FS corresponding to another new set of random variables can be readily calculated. This facilitates the evaluation of failure probability in the reliability analysis of earth dam slope stability, as discussed in the next subsection.

9.4.2 *Evaluation of the Failure Probability Using XGBoost*

In geotechnical reliability analysis, the failure probability P_f (or reliability index) is frequently used to measure the safety of geotechnical structures from a probabilistic perspective. However, it is difficult to solve the P_f analytically by integrating the performance function over the failure zone, because the performance function is usually implicit and highly nonlinear in geotechnical engineering. In such a case, it is preferable to calculate the P_f approximately by MCS due to its conceptual simplicity and easy to use for geotechnical engineers. Based on a total number of N_{MC} random field samples generated from the midpoint method, the FS_i $(i = 1,\ 2,\ \ldots,\ N_{\mathrm{MC}})$ corresponding to each random field realization can be evaluated by the established XGBoost. Then, the P_f is expressed as:

$$P_f = \frac{1}{N_{\mathrm{MC}}} \sum_{i=1}^{N_{\mathrm{MC}}} I[\mathrm{FS}_i < 1] \tag{9.11}$$

where $I[\cdot]$ is an indicator function for judging the earth dam slope failure occurs or not. If $\mathrm{FS}_i < 1$, $I[\cdot] = 1$; otherwise, $I[\cdot] = 0$. As indicated by Eq. (9.11), it is a cumbersome work to calculate P_f which involves a huge number of deterministic slope stability analyses. In such a case, the established XGBoost is employed to predict the FS_i in this study for facilitating the reliability analysis of earth dam slope stability. The detailed implementation procedures involved in the proposed approach are summarized in the next section.

9.5 Implementation Procedure

To facilitate the understanding and application of the proposed XGBoost-based reliability analysis approach in engineering practice for geotechnical engineers, its implement procedures outlined in Fig. 9.1 can be divided into four parts, including random field modeling, deterministic analysis, XGBoost model calibration and evaluation of P_f using the established XGBoost. Firstly, generate a total of N_{MC}^T random field samples of soil properties using the midpoint method based on the predetermined information, such as earth dam geometry and statistics of soil properties (e.g., means and standard deviations). Then, deterministic seepage and slope stability analyses are conducted to calculate the FS of each random field sample, leading to a total of N_{MC}^T values of FS. In this study, the commercial software Geostudio is applied to perform the seepage and slope stability analyses of earth dam. Subsequently, divide these N_{MC}^T FS values associated with their inputs into training dataset and validation dataset for constructing the XGBoost model. Thereafter, the k-fold cross-validation technique is employed to evaluate its accuracy. Consider, for example, a total of N_{MC}^P random field samples simulated from the midpoint method are used for predicting the earth dam slope failure probability. Finally, the N_{MC}^{P} FS values and the corresponding failure probability P_f can be readily evaluated by virtue of the established XGBoost model.

To facilitate the practical application of the proposed approach, these implementation procedures can be programmed as user functions or toolboxes. These packaged user functions or toolboxes are then used as a "black box" for geotechnical engineers to carry out reliability analysis task in engineering practice. For illustration, the proposed approach is applied to the reliability analysis of Ashigong earth dam in the next section.

9.6 Application to Ashigong Earth Dam Slope

The Ashigong earth dam is situated on the Yellow River in Guide County of Qinghai Province, China. The normal storage water level is 2198 m, and the maximum reservoir capacity is 6836×10^4 m^3 with an installed electricity capacity of 180 MW. According to the existing standards of rank classification of hydropower projects, the Ashigong earth dam belongs to the medium-sized third-class project. As shown in Fig. 9.2, the earth dam is mainly constituted by dam body, inclined loam core, concrete cut-off wall, and dam foundation. The dam body of Ashigong is filled with sandy gravel, and the crest elevation and ground surface elevation are 2202 m and 2179 m, respectively. A 50 m thick dam foundation (i.e., elevation ranges from 2129 to 2179 m) is considered in this example, which is comprised of sandy gravel layer, silty clay layer, silty fine sand interlayer, silty fine sand layer, silty clay layer, gravel layer, and argillaceous siltstone layer according to the geological survey report.

To simulate the seepage through Ashigong earth dam, hydraulic boundary conditions need to be predefined. The initial upstream and downstream reservoir water levels are 2198 m and 2181 m, respectively. Then, two constant total heads corresponding to the upstream and downstream reservoir water levels are applied to the earth dam below the water level. Besides, a zero flux condition is assigned to the left and right boundaries of the dam foundation. By assuming that the Ashigong earth dam is underlain by an impermeable layer, a zero flux condition is also applied to the bottom boundary. Table 9.1 tabulates the mean values of soil properties of the earth dam materials, which are adapted from the geological survey report. Besides these mean values, Table 9.1 also summarized the COVs and SOFs of soil properties, which are adapted from the previous literature (e.g., Li et al. 2015; Liu et al. 2017) due to the lack of measurement data. Based on these parameters, the midpoint method can be used to generate stationary random field samples of soil properties.

In this example, saturated hydraulic conductivity k_s and effective friction angle φ' are regarded as lognormal random fields instead of normal random fields, so as to avoid the possible negative cases. Due to the lack of SWCC test data for sandy gravel filler material, the VGM model and its corresponding fitting parameters (i.e., α_{vg} and n_{vg}) obtained from literature (e.g., Sillers and Fredlund 2001; Cho 2012) are used for illustration. Compared with the saturated volumetric water content θ_s measured

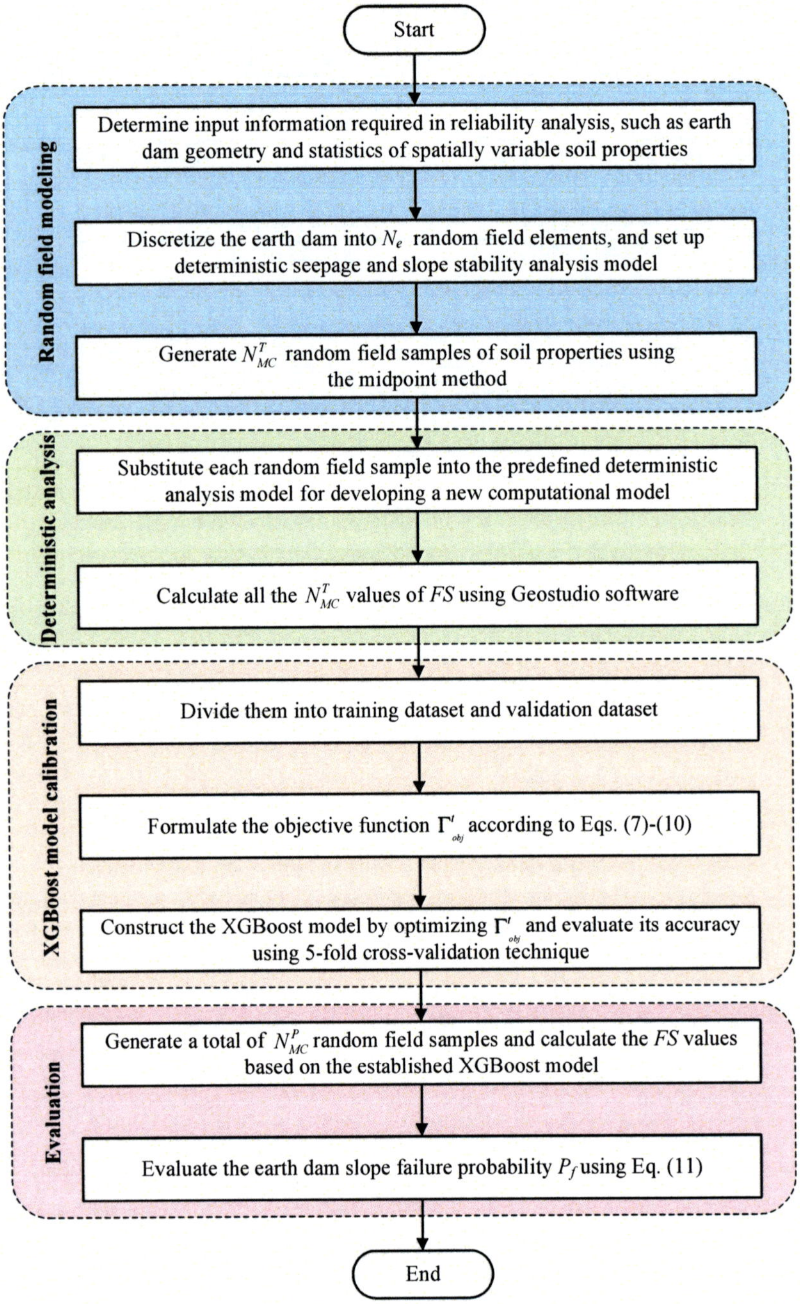

Fig. 9.1 Flow chart for the XGBoost-based reliability analysis approach

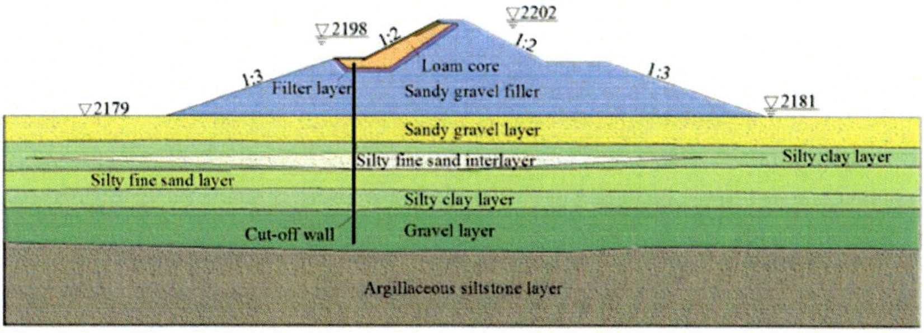

Fig. 9.2 Location and typical section of Ashigong earth dam

from tests, the residual volumetric water content θ_r is generally unavailable and is frequently taken as zero (e.g., Phoon et al. 2010).

Based on the mean values of soil properties summarized in Table 9.1, the SEEP/W module and SLOPE/W module contained in the commercial software Geostudio are employed to perform steady state seepage and slope stability analyses of Ashigong earth dam, respectively. Firstly, the SEEP/W module is applied to analyze the hydrological behavior of earth dam considering steady state seepage of reservoir water, such as the phreatic surface (i.e., the surface with zero pressure head) plotted in Fig. 9.3. As indicated by the extended Mohr–Coulomb equation, matric suction (i.e., pore-water pressure minus pore-air pressure) and volumetric water content are needed to evaluate the shear strength of unsaturated soils (Vanapalli et al. 1996). Then, the seepage results obtained from SEEP/W are incorporated into the SLOPE/W to calculate the FS value. In this example, the FS of Ashigong earth dam is obtained as 1.198 using the Morgenstern–Price method, as shown in Fig. 9.3. For reference, Fig. 9.3 also plots the location of critical slip surface.

By the random field modeling technique (i.e., the midpoint method in Sect. 9.3), a large number of random field samples of soil properties can be generated. These random field samples are then used as inputs in Geostudio for evaluating the corresponding values of FS by repetitively performing steady state seepage and slope stability analyses. Note that although the stationary random fields are considered in this example due to the lack of measurement data, other advanced random field modeling techniques of interest can also be applied to depict the spatially varying soil properties as the site-specific measurement data are available (e.g., BCS-KL method). Based on these random field samples and the corresponding FS values, XGBoost is used to reconstruct the relationship between the spatially variable soil properties and FS for Ashigong earth dam.

Table 9.1 Statistical information of soil properties for Ashigong earth dam

Soil type	Parameter	Mean*	COV#	Distribution#	SOF#	
					δ_h	δ_v
Sandy gravel filler	k_s (m/s)	5.00×10^{-5}	0.5	Lognormal	40	4
	φ' (°)	29	0.2			
	c' (kPa)	0	—			
	γ (kN/m^3)	22				
	α_{vg} (kPa)	5				
	n_{vg}	2.68				
	θ_s	0.39				
	θ_r	0				
Sandy gravel layer	k_s (m/s)	3.20×10^{-5}	0.5	Lognormal	40	4
	φ' (°)	27	0.2			
	c' (kPa)	0	—			
	γ (kN/m^3)	19.9				
Silty clay layer	k_s (m/s)	5.00×10^{-7}	0.5	Lognormal	40	4
	φ' (°)	11	0.2			
	c' (kPa)	16	—			
	γ (kN/m^3)	18				
Silty fine sand layer	k_s (m/s)	5.00×10^{-6}	0.5	Lognormal	40	4
	φ' (°)	21	0.2			
	c' (kPa)	0	—			
	γ (kN/m^3)	18.8				
Gravel layer	k_s (m/s)	3.20×10^{-5}	0.5	Lognormal	40	4
	φ' (°)	30	0.2			
	c' (kPa)	0	–			
	γ (kN/m^3)	21.2				

Note k_s is the saturated hydraulic conductivity; c' is the effective cohesion; φ' is the effective friction angle; γ is the bulk unit weight; α_{vg} and n_{vg} are the fitting parameters of VGM model; θ_s and θ_r are the saturated volumetric water content and residual volumetric water content, respectively; Symbol "*" denotes the item adapted from geological survey report; Symbol "#" denotes the item that is referred to Li et al. (2015) and Liu et al. (2017); Symbol "–" denotes the not applicable item

9.6.1 Construction of XGBoost Model

A certain number (i.e., N_{MC}^T) of training samples are essential prerequisites for establishing the XGBoost model. However, there is no guideline on determining the number of training samples in the present study. Although a large number of training samples may improve the model performance of XGBoost, it is a non-trivial task to prepare these training samples in geological reliability analysis which requires extensive computational efforts for conducting the deterministic analysis

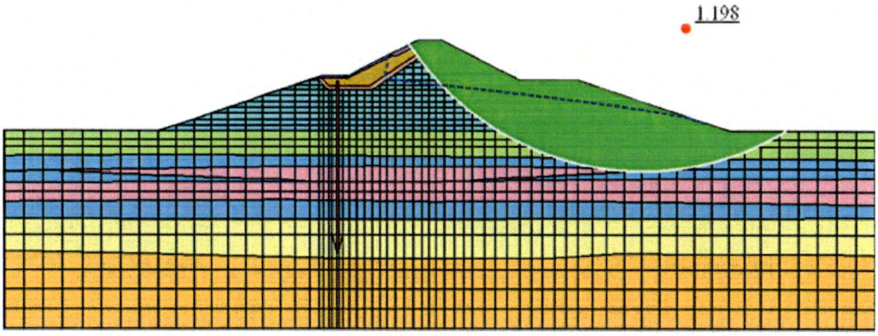

Fig. 9.3 Slope stability results of Ashigong earth dam obtained from deterministic analysis

repeatedly. For illustration, the N_{MC}^{T} equals the number of random field elements (i.e., $N_{\mathrm{MC}}^{T} = 2198$) in this example. Then, a total of 2198 random field samples are generated by the midpoint method and the corresponding values of FS are evaluated from Geostudio software. These random field samples and FS values constitute the training dataset for constructing the XGBoost model. It is necessary to prescribe the number of features before the construction process, which influences the model complexity and model performance and can be determined by feature engineering (Fan et al. 2019). In this example, a total of 1100 features are used in this example. The k-fold cross-validation technique is usually used to evaluate the model performance, and previous researches indicated that fivefold cross-validation is practically applicable and able to provide convincing results (e.g., Rodríguez et al. 2010; Wong 2015). Thus, the fivefold cross-validation is employed in this example. To facilitate the understanding of the modeling process of XGBoost, Fig. 9.4 outlines a schematic diagram for constructing the XGBoost model.

Based on the 2198 training samples, an XGBoost regression model containing a total of 850 trees (i.e., $K = 850$) is constructed after continuous debugging. Figure 9.5 shows the results of fivefold cross-validation obtained from the XGBoost regression model. As observed from Fig. 9.5a, all the values of coefficient of determination R^2 for five training datasets and validation datasets are larger than 0.90 and 0.80, respectively. In addition, the corresponding failure probabilities for all the training datasets and validation datasets are also plotted in Fig. 9.5b. Compared with the satisfying results of R^2 shown in Fig. 9.5a, there exists an undesirable discrepancy between the P_f obtained from the validation datasets and the true value (i.e., 0.0464) evaluated from the MCS. This apparent discrepancy may be attributed to possible biases in the estimation of FS values less than the threshold (i.e., 1), whereas these failure samples are of great concern in geotechnical reliability analysis for calculating the failure probability as indicated by Eq. (9.11). To validate this guess, Fig. 9.6 compares the FS values obtained from the established XGBoost regression model and deterministic analysis for 2198 training samples and 10,000 testing samples. In general, the predicted FS values obtained from the established XGBoost regression model

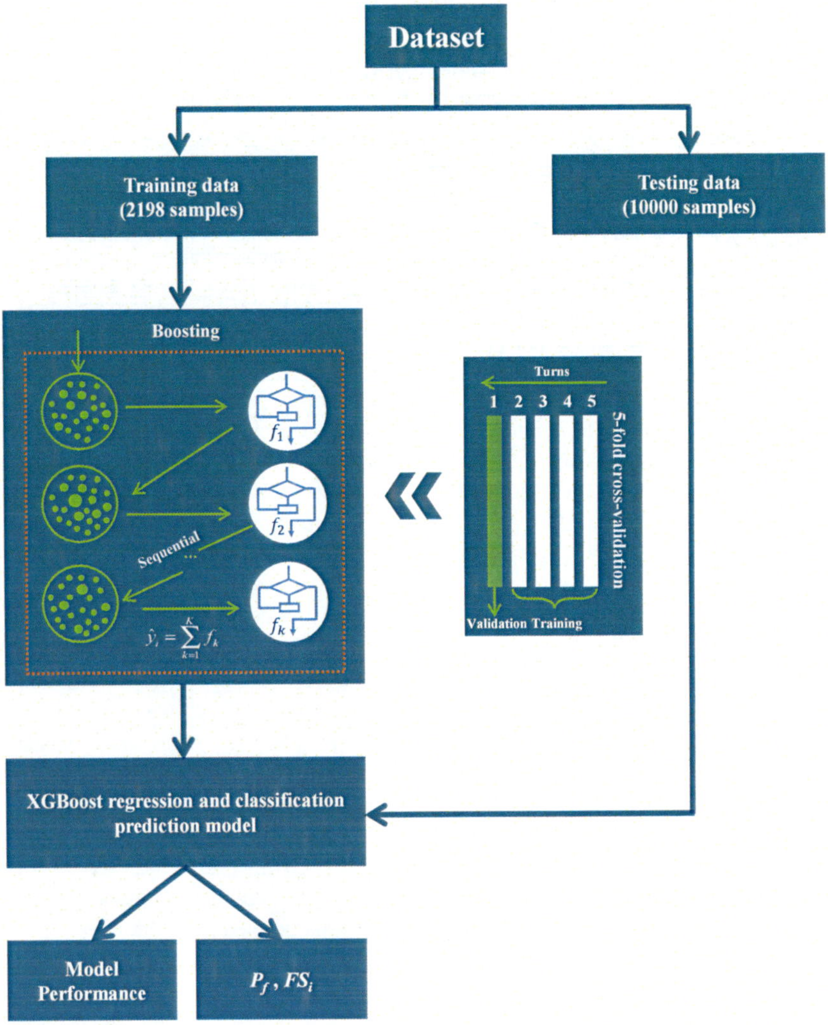

Fig. 9.4 Schematic diagram of XGBoost modeling

agree well with those evaluated from deterministic analysis for both the training and testing samples. The corresponding R^2 values of these two scenarios are as high as 0.966 and 0.854, respectively. Nonetheless, for the failure samples (i.e., FS < 1) of interest, it is evident that the predicted values are larger than those calculated from deterministic analysis, as shown in Fig. 9.6b. This implies that the established XGBoost regression model may be unable to capture the failure samples well in this example. It is well recognized that the failure probabilities of geotechnical structures are usually at a small level, which can be regarded as an imbalanced data classification problem from a machine learning perspective. In such a case, the XGBoost classification model may be superior to the regression model in predicting the failure probability in this example. For XGBoost classification model, the values of FS are categorized into two classes according to the value less than 1 or not. By adjusting the predictive weight of imbalanced data, the failure samples can be predicted precisely and rationally. Based on the established XGBoost classification model, the fivefold cross-validation is applied to evaluate its accuracy.

Figure 9.7 plots the results of fivefold cross-validation obtained from XGBoost classification model. To assess the performance of XGBoost classification model, the frequently used indexes, namely accuracy and area under the receiver operating characteristic curve (AUC of ROC), are used in this example. Accuracy denotes the proportion of correctly predicted samples in the total number of samples. As shown in Fig. 9.7a, both the accuracy values of five training datasets and validation datasets are greater than 0.95. For further verification, the other index of AUC is also utilized, which has gained increasing popularity in geotechnical engineering (Hu and Liu 2019). The ROC curve represents the relationship between the true positive rate and false positive rate, which is generally used for examining the generalization ability of the model. For qualitatively evaluate the generalization performance, the AUC index describing the area under the ROC curve is frequently used. Generally, the AUC value ranges from 0.5 to 1, and a higher AUC value implies the model with a better model performance. Figure 9.7b also plots the ROC curves corresponding to five validation datasets. It can be observed that all the five ROC curves have relatively higher values of AUC, indicating that the XGBoost classification model performs well in this example. Furthermore, Fig. 9.7c compares the failure probabilities obtained from the five training datasets and validation datasets. It is shown that the P_f of five training and validation datasets are generally in good agreement with the true value (i.e., 0.0464). This validates the excellent performance of XGBoost classification model in the reliability analysis. Finally, the established XGBoost classification model is used to predict the P_f of 10,000 testing samples, the ROC curve associated with the AUC value is plotted in Fig. 9.8. Likewise, a relatively higher AUC value can also be observed. Besides, the P_f predicted from the established XGBoost classification model agrees well with that calculated from the MCS. This again verifies the proposed XGBoost-based reliability analysis approach, which opens up a possibility of facilitating the geotechnical reliability analysis with the aid of advanced machine learning algorithms.

Besides the model accuracy, computational cost is also a primary concern in geotechnical applications. In this study, the execution time is used to measure the computational cost of the proposed approach and the MCS method. The XGBoost model should be first calibrated on a total of 2198 training samples obtained from repeated deterministic analyses, which takes about 3 h and 20 min on a personal desktop computer with 16 GB RAM and 12 Intel Core i7 CPU clocked at 3.70 GHz. Subsequently, it requires about 15 s by applying the XGBoost model for predicting the failure probability of 10,000 testing samples. It can be observed that the execution time of model prediction (i.e., about 15 s) is marginal when compared with the time request of model calibration (i.e., about 3 h and 20 min). The total execution time of the proposed approach is about 3 h and 21 min, which contains both the time

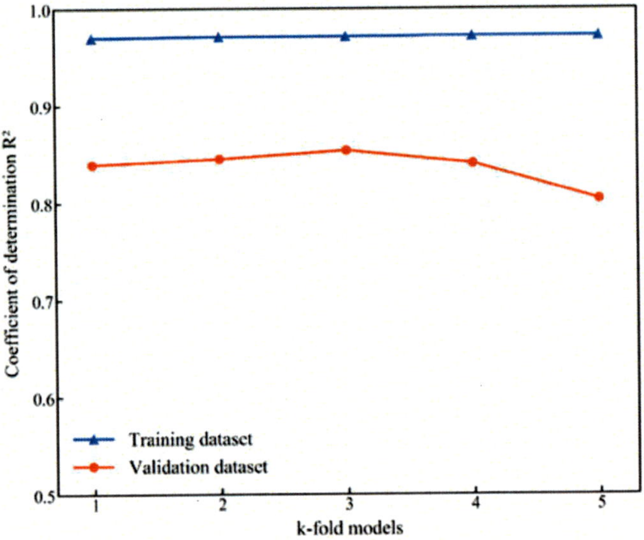

(a) Coefficient of determination R^2

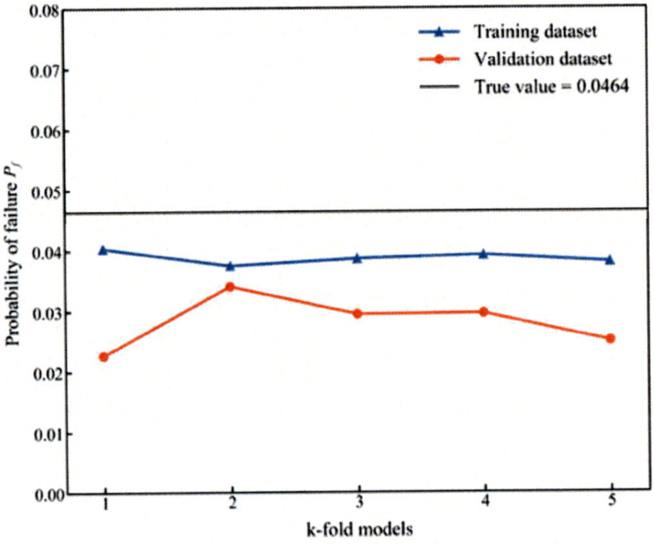

(b) Failure probability P_f

Fig. 9.5 Results of fivefold cross-validation obtained from XGBoost regression model

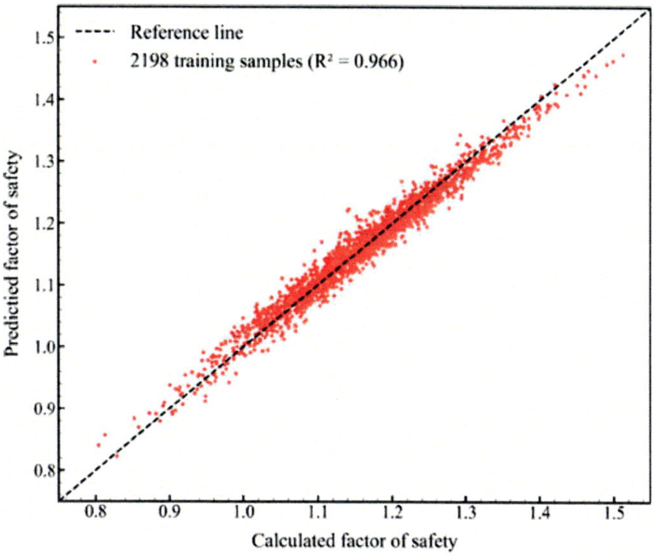

(a) Training samples

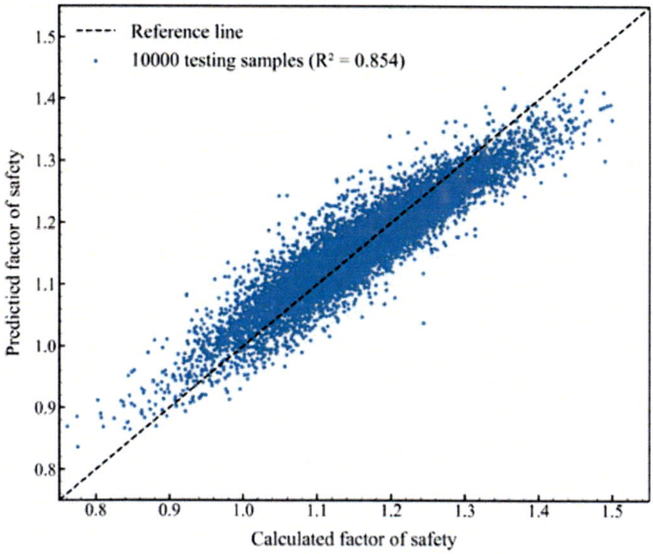

(b) Testing samples

Fig. 9.6 Validation of XGBoost regression model using training and testing samples

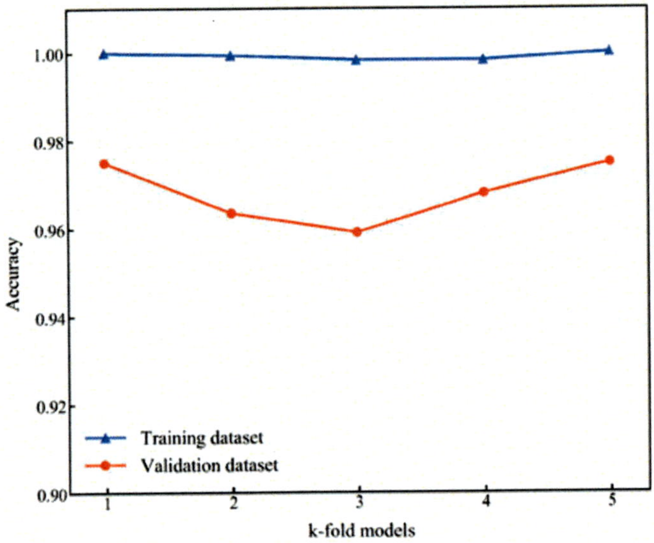

(a) Accuracy

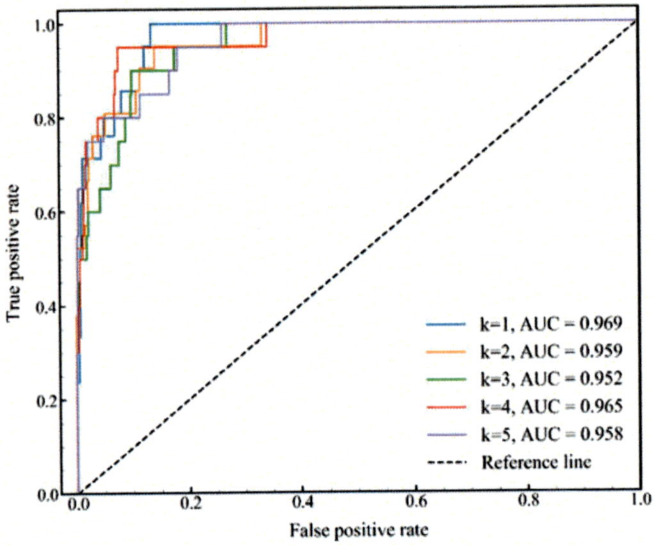

(b) ROC curves and the AUC values of validation dataset

Fig. 9.7 Results of fivefold cross-validation obtained from XGBoost classification model

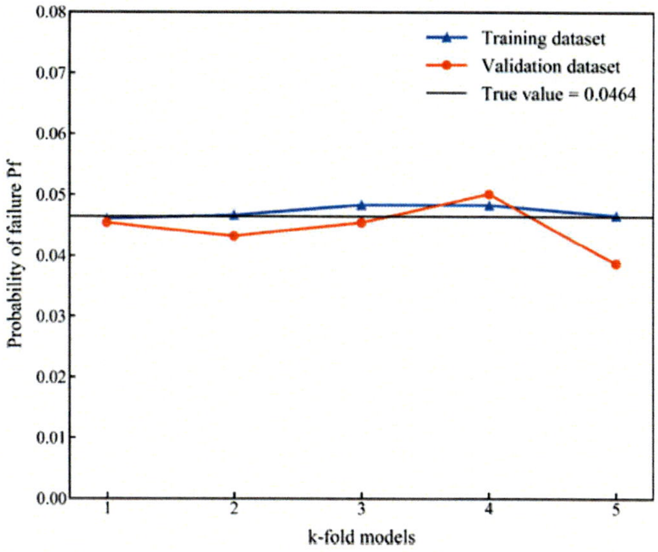

(c) Failure probability

Fig. 9.7 (continued)

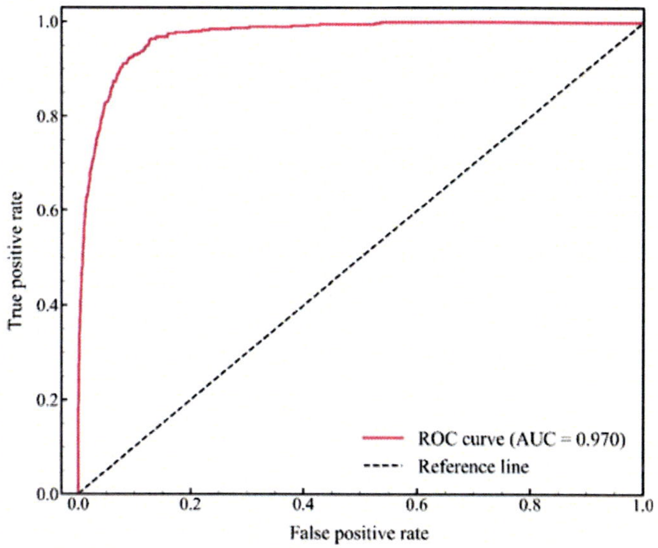

Fig. 9.8 ROC curve and the corresponding value of AUC

for XGBoost model calibration and prediction. For the direct MCS method, a total of 10,000 deterministic analyses are performed repeatedly through the Geostudio software, which takes about 14 h and 5 min on the same desktop computer. In general, the execution time of the XGBoost is significantly less than that of the direct MCS method in this example. This reveals the feasibility of the proposed approach in practical applications. Based on the proposed approach, a parametric sensitivity analysis is conducted to investigate the influences of spatially variable soil properties on the earth dam slope failure probability systematically, as discussed in the following subsections.

9.6.2 *Effect of COV on the Earth Dam Slope Failure Probability*

To investigate the effects of coefficients of variation of k_s and φ' (i.e., COV_{k_s} and $\mathrm{COV}_{\varphi'}$) on the failure probability of the Ashigong earth dam slope, a parametric sensitivity analysis is conducted in this section. Inspired by the values of COV_{k_s} and $\mathrm{COV}_{\varphi'}$ reported in the literature (e.g., Srivastava et al. 2010; Cho 2012; Li et al. 2015), the ranges of COV_{k_s} and $\mathrm{COV}_{\varphi'}$ are set as [0.3, 1.0] and [0.05, 0.20] in this study, respectively. In the parametric sensitivity analysis, the parameter values provided in Table 9.1 are regarded as the baseline case for reference. Based on the newly prescribed value of COV and the remaining statistical information (e.g., means and scale of fluctuations) provided in the baseline case, a total of 10,000 (i.e., N_{MC}^{P}) random field samples of k_s and φ' are simulated by midpoint method. Then, the corresponding failure probability P_f can be readily calculated with the aid of the proposed XGBoost-based approach.

Figure 9.9 plots the variation of P_f with different values of COV_{k_s} and $\mathrm{COV}_{\varphi'}$. It is observed that the P_f of earth dam slope increases with the increase of COV_{k_s} (see blue line in Fig. 9.9a), indicating that the earth dam slope failure probability is affected by the variations of COV_{k_s}. In addition, the P_f of earth dam slope increases dramatically as the $\mathrm{COV}_{\varphi'}$ increases (see blue line in Fig. 9.9b). Particularly, when $\mathrm{COV}_{\varphi'}$ increases from 0.10 to 0.20, the corresponding value of P_f increases sharply from about 0.0067 to 0.0452. This indicates that the coefficient of variation of φ' affects the earth dam slope failure probability significantly. Besides the single exponential ACF (i.e., SECF) used in the baseline case, other ACFs can also be applied to characterize the spatial variability of k_s and φ'. For illustration, three other ACFs are selected in this sensitivity analysis for investigating their effects on the Ashigong earth dam slope failure probability, which are second-order Markov, cosine exponential and binary noise ACFs. For simplification, these three ACFs are abbreviated as SMCF, CECF, and BNCF in this study, respectively.

Figure 9.9 also shows the variation of P_f with the coefficients of variation of k_s and φ' for different ACFs. As expected, the selection of ACF exerts a significant influence on the earth dam slope failure probability. Among the four ACFs, the P_f

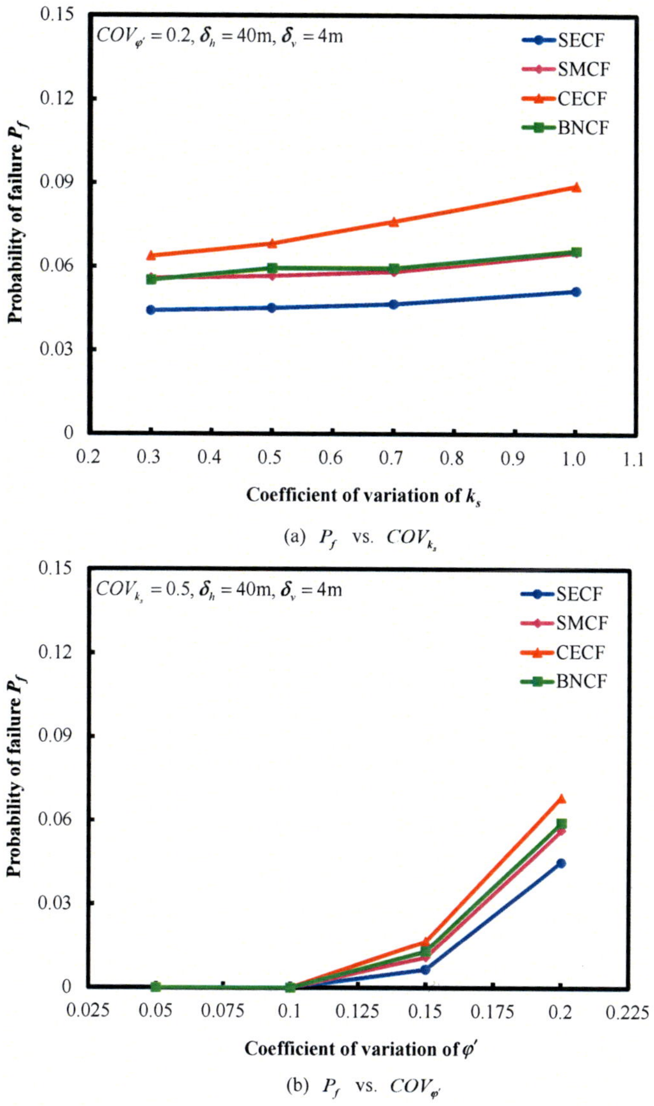

Fig. 9.9 Variation of P_f with the coefficients of variation of k_s and φ'

obtained from CECF is obviously greater than that calculated from the other three ACFs (see red line in Fig. 9.9). In particular, the SECF gives the smallest value of P_f among them (see blue line in Fig. 9.9), indicating that the commonly used SECF tends to underestimate the earth dam slope failure probability and provide an unconservative estimate in the reliability analysis of earth dam slope stability. This further validates the findings observed in Li et al. (2015), where the slope failure probability is highly dependent on the selection of ACF. Thus, it is necessary to determine an appropriate ACF in practical geotechnical reliability analysis. In addition to the ACF, the horizontal and vertical SOFs (i.e., δ_h and δ_v) may also influence the earth dam slope failure probability, as discussed in the following subsection.

9.6.2.1 Effect of SOF on the Earth Dam Slope Failure Probability

To further explore the influences of horizontal and vertical SOFs (i.e., δ_h and δ_v) of k_s and φ' on the failure probability of the Ashigong earth dam slope, different values of δ_h and δ_v are adopted to perform parametric sensitivity analysis. Following Cho (2012) and Li et al. (2015), the typical ranges of δ_h and δ_v are selected as [10 m, 60 m] and [1 m, 6 m] in this study, respectively. Then, the P_f corresponding to the newly generated random field samples can be conveniently evaluated using the proposed XGBoost-based approach. Figure 9.10 shows the variation of P_f with different values of δ_h and δ_v. It can be observed that the P_f of earth dam slope increases monotonically with the increase of δ_h and δ_v (see blue line in Fig. 9.10), which is consistent with that reported in the literature (e.g., Le et al. 2015; Liu et al. 2019). It can be deduced that the P_f may be overestimated when the spatial variability of k_s and φ' is ignored and they are simply regarded as random variables (i.e., $\delta_h \rightarrow \infty$ and $\delta_v \rightarrow \infty$) in the reliability analysis of earth dam slope stability.

Moreover, Fig. 9.10 also compares the results of P_f obtained from the four different ACFs (i.e., SECF, SMCF, CECF, and BNCF). It can be seen that all the values of P_f obtained from the four ACFs increase as δ_h and δ_v become larger. Similarly, the CECF gives the greatest value of P_f among the four ACFs (see red line in Fig. 9.10), implying that the CECF tends to provide a conservative estimate. In contrast, the P_f obtained from SECF is obviously less than that calculated from the other three ACFs (see blue line in Fig. 9.10). This reveals that the commonly used SECF may underestimate the failure probability in the reliability analysis of earth dam slope stability. Note that although the earth dam slope failure probability is evidently affected by ACF and its corresponding horizontal and vertical SOFs, it is difficult to determine the suitable ACF and SOFs in geotechnical engineering practice due to the limited measured data (e.g., Cao and Wang 2014). Recently, a novel random field generator called Bayesian compressing sampling (BCS) and Karhunen-Loève (KL) generator has gained increasing attention in geotechnical and geological engineering (e.g., Wang et al. 2018c; Hu et al. 2019), which combines the advantages of the BCS and KL expansion. It provides an effective means to generate the random field samples directly from the limited measurement data at a specific site, thereby avoiding the prior assumption on the function form of the autocorrelation structure.

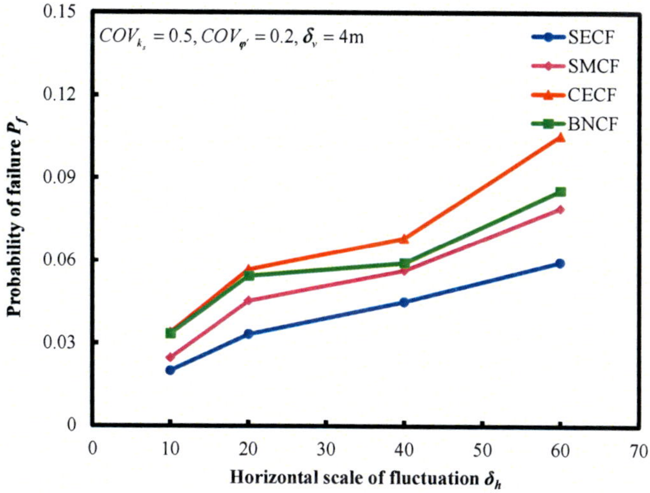

(a) P_f vs. δ_h

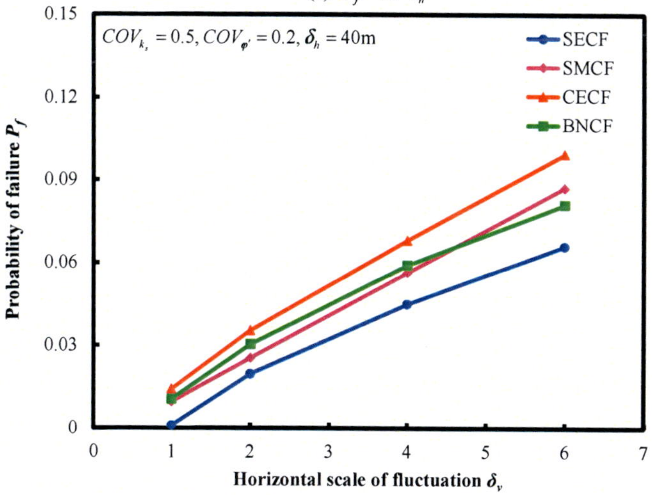

(b) P_f vs. δ_v

Fig. 9.10 Variation of P_f with the coefficients of variation of δ_h and δ_v

9.7 Summary and Conclusions

This chapter developed a XGBoost-based reliability analysis approach for earth dam slope stability analysis. The proposed approach introduces an advanced machine learning algorithm called XGBoost to evaluate the earth dam slope failure probability in spatially variable soils, avoiding the extensive computational efforts required in the direct Monte Carlo simulation. Moreover, the proposed approach is decoupled from the deterministic seepage and slope stability analysis, so that the XGBoost-based reliability analysis approach can conveniently proceed as an extension of deterministic analysis in a non-intrusive manner. It not only facilitates the reliability analysis of earth dam slope stability in practical applications, but also allows geotechnical engineers concentrating more on the familiar deterministic analysis. The proposed approach is applied to the reliability analysis of Ashigong earth dam located in Qinghai Province of China, and a parametric sensitivity analysis is performed to investigate the influences of spatially variable soil properties on the earth dam slope failure probability. The following conclusions can be drawn from this chapter:

(1) The proposed XGBoost-based reliability analysis approach provides an effective way to evaluate the earth dam slope failure probability considering the spatial variability of soil properties. It is shown that the XGBoost classification model performs well in the fivefold cross-validation. Furthermore, the proposed approach is able to predict the failure probability of earth dam slope in spatially variable soils with reasonable accuracy and efficiency. It opens up a possibility of integrating advanced machine learning algorithms into geotechnical reliability analysis for facilitating the reliability analysis of geotechnical structures in practical applications.
(2) The earth dam slope failure probability is significantly affected by the COVs of soil properties (i.e., saturated hydraulic conductivity and effective friction angle). It is observed that the failure probability of earth dam slope increases with the increase of COV. Besides, both the horizontal and vertical SOFs of soil properties considerably affect the earth dam slope failure probability. It can be deduced that the failure probability may be overestimated when the spatial variability of soil properties is neglected and they are simply regarded as random variables in the reliability analysis of earth dam slope stability.
(3) The selection of ACF has a significant influence on the earth dam slope failure probability. Among the four different ACFs, the SECF gives the smallest value of failure probability, implying that the commonly used SECF tends to underestimate the earth dam slope failure probability and provide an unconservative estimate in the reliability analysis of earth dam slope stability. In general, the earth dam slope failure probability is affected by the spatial variability of soil properties.

References

Ahmed AA (2009) Stochastic analysis of free surface flow through earth dams. Comput Geotech 36(7):1186–1190. https://doi.org/10.1016/j.compgeo.2009.05.005

Au SK, Beck JL (2001) Estimation of small failure probabilities in high dimensions by subset simulation. Probab Eng Mech 16(4):263–277. https://doi.org/10.1016/S0266-8920(01)00019-4

Au SK, Wang Y (2014) Engineering risk assessment with subset simulation. Wiley, Singapore

Babu GLS, Srivastava A (2010) Reliability analysis of earth dams. J Geotech Geoenviron Eng 136(7):995–998. https://doi.org/10.1061/(ASCE)GT.1943-5606.0000313

Cao ZJ, Wang Y (2014) Bayesian model comparison and selection of spatial correlation functions for soil parameters. Struct Saf 49:10–17. https://doi.org/10.1016/j.strusafe.2013.06.003

Chen TQ, Guestrin C (2016) XGBoost: a scalable tree boosting system. In: Proceedings of the 22nd ACM SIGKDD international conference on knowledge discovery and data mining, 785–794. https://doi.org/10.1145/2939672.2939785

Childs EC, Collis-George N (1950) The permeability of porous materials. Proc Soc London Ser a Math Phys Sci 201(1066):392–405. https://doi.org/10.1098/rspa.1950.0068

Cho SE (2010) Probabilistic assessment of slope stability that considers the spatial variability of soil properties. J Geotech Geoenviron Eng 136(7):975–984. https://doi.org/10.1061/(ASCE)GT.1943-5606.0000309

Cho SE (2012) Probabilistic analysis of seepage that considers the spatial variability of permeability for an embankment on soil foundation. Eng Geol 133–134:30–39. https://doi.org/10.1016/j.enggeo.2012.02.013

Fan C, Sun Y, Zhao Y, Song M, Wang J (2019) Deep learning-based feature engineering methods for improved building energy prediction. Appl Energy 240:35–45. https://doi.org/10.1016/j.apenergy.2019.02.052

Fenton GA, Griffiths DV (1996) Statistics of free surface flow through stochastic earth dam. J Geotech Eng 122(6):427–436. https://doi.org/10.1061/(ASCE)0733-9410(1996)122:6(427)

Fenton GA, Griffiths DV (1997) Extreme hydraulic gradient statistics in stochastic earth dam. J Geotech Geoenviron Eng 123(11):995–1000. https://doi.org/10.1061/(ASCE)1090-0241(1997)123:11(995)

Fredlund DG, Rahardjo H, Fredlund MD (2012) Unsaturated soil mechanics in engineering practice. Wiley, Hoboken, NJ

GEO-SLOPE International, Ltd. (2012) Geostudio. https://www.geoslope.com/

Gui SX, Zhang RD, Turner JP, Xue XZ (2000) Probabilistic slope stability analysis with stochastic soil hydraulic conductivity. J Geotech Geoenviron Eng 126(1):1–9. https://doi.org/10.1061/(ASCE)1090-0241(2000)126:1(1)

Hicks MA, Li YJ (2018) Influence of length effect on embankment slope reliability in 3D. Int J Numer Anal Methods Geomech 42(7):891–915. https://doi.org/10.1002/nag.2766

Hu JL, Liu HB (2019) Bayesian network models for probabilistic evaluation of earthquake-induced liquefaction based on CPT and Vs databases. Eng Geol 254:76–88. https://doi.org/10.1016/j.enggeo.2019.04.003

Hu Y, Zhao TY, Wang Y, Choi C, Ng CWW (2019) Direct simulation of two-dimensional isotropic or anisotropic random field from sparse measurement using Bayesian compressive sampling. Stoch Environ Res Risk Assess 33(8–9):1477–1496. https://doi.org/10.1007/s00477-019-01718-7

Huang HW, Wen SC, Zhang J, Chen FY, Martin JR, Wang H (2018) Reliability analysis of slope stability under seismic condition during a given exposure time. Landslides 15(11):2303–2313. https://doi.org/10.1007/s10346-018-1050-9

Huang XX, Chen JQ, Zhu HP (2016) Assessing small failure probabilities by AK-SS: an active learning method combining Kriging and Subset simulation. Struct Saf 59:86–95. https://doi.org/10.1016/j.strusafe.2015.12.003

Ji J, Zhang CS, Gao YF, Kodikara J (2018) Effect of 2D spatial variability on slope reliability: a simplified FORM analysis. Geosci Front 9(6):1631–1638. https://doi.org/10.1016/j.gsf.2017.08.004

Khalilzad M, Gabr MA, Ellen M (2015) Assessment of remedial measures to reduce exceedance probability of performance limit states in embankment dams. Comput Geotech 67:213–222. https://doi.org/10.1016/j.compgeo.2015.02.010

Le TMH, Gallipoli D, Sanchez M, Wheeler SJ (2012) Stochastic analysis of unsaturated seepage through randomly heterogeneous earth embankments. Int J Numer Anal Methods Geomech 36(8):1056–1076

Le TMH, Gallipoli D, Sánchez M, Wheeler S (2015) Stability and failure mass of unsaturated heterogeneous slopes. Can Geotech J 52(11):1747–1761. https://doi.org/10.1139/cgj-2014-0190

Leong EC, Rahardjo H (1997) Permeability functions for unsaturated soils. J Geotech Geoenviron Eng 123(12):1118–1126. https://doi.org/10.1061/(ASCE)1090-0241(1997)123:12(1118)

Li DQ, Jiang SH, Cao ZJ, Zhou W, Zhou CB, Zhang LM (2015) A multiple response-surface method for slope reliability analysis considering spatial variability of soil properties. Eng Geol 187:60–72. https://doi.org/10.1016/j.enggeo.2014.12.003

Li DQ, Wang L, Cao ZJ, Qi XH (2019) Reliability analysis of unsaturated slope stability considering SWCC model selection and parameter uncertainties. Eng Geol 260:105207. https://doi.org/10.1016/j.enggeo.2019.105207

Li DQ, Xiao T, Cao ZJ, Zhou CB, Zhang LM (2016a) Enhancement of random finite element method in reliability analysis and risk assessment of soil slopes using Subset Simulation. Landslides 13(2):293–303. https://doi.org/10.1007/s10346-015-0569-2

Li DQ, Zheng D, Cao ZJ, Tang XS, Phoon KK (2016b) Response surface methods for slope reliability analysis: review and comparison. Eng Geol 203:3–14. https://doi.org/10.1016/j.enggeo.2015.09.003

Li XY, Zhang LM, Zhang S (2018) Efficient Bayesian networks for slope safety evaluation with large quantity monitoring information. Geosci Front 9(6):1679–1687. https://doi.org/10.1016/j.gsf.2017.09.009

Liu LL, Cheng YM (2016) Efficient system reliability analysis of soil slopes using multivariate adaptive regression splines-based Monte Carlo simulation. Comput Geotech 79:41–54. https://doi.org/10.1016/j.compgeo.2016.05.001

Liu LL, Cheng YM, Jiang SH, Zhang SH, Wang XM, Wu ZH (2017) Effects of spatial autocorrelation structure of permeability on seepage through an embankment on a soil foundation. Comput Geotech 87:62–75. https://doi.org/10.1016/j.compgeo.2017.02.007

Liu LL, Zhang SH, Cheng YM, Liang L (2019) Advanced reliability analysis of slopes in spatially variable soils using multivariate adaptive regression splines. Geosci Front 10(2):671–682. https://doi.org/10.1016/j.gsf.2018.03.013

Liu Y, Zhang WG, Zhang L, Zhu ZR, Hu J, Wei H (2018) Probabilistic stability analyses of undrained slopes by 3D random fields and finite element methods. Geosci Front 9(6):1657–1664. https://doi.org/10.1016/j.gsf.2017.09.003

Montoya-Noguera S, Zhao TY, Hu Y, Wang Y, Phoon KK (2019) Simulation of non-stationary non-Gaussian random fields from sparse measurements using Bayesian compressive sampling and Karhunen-Loève expansion. Struct Saf 79:66–79. https://doi.org/10.1016/j.strusafe.2019.03.006

Mualem Y (1976) A new model for predicting the hydraulic conductivity of unsaturated porous media. Water Resour Res 12:513–522. https://doi.org/10.1029/WR012i003p00513

Phoon KK, Kulhawy FH (1999) Characterization of geotechnical variability. Can Geotech J 36(4):612–624. https://doi.org/10.1139/t99-038

Phoon KK, Santoso A, Quek ST (2010) Probabilistic analysis of soil-water characteristic curves. J. Geotech. Geoenviron. Eng. 136:445–455. https://doi.org/10.1061/(ASCE)GT.1943-5606.0000222

Richards LA (1931) Capillary conduction of liquids through porous mediums. Physics 1(5):318–333. https://doi.org/10.1063/1.1745010

Rodríguez JD, Pérez A, Lozano JA (2010) Sensitivity analysis of k-fold cross validation in prediction error estimation. IEEE Trans Pattern Anal Mach Intell 32(3):569–575. https://doi.org/10.1109/TPAMI.2009.187

Santoso AM, Phoon KK, Quek ST (2011) Effects of soil spatial variability on rainfall-induced landslides. Comput Struct 89:893–900. https://doi.org/10.1016/j.compstruc.2011.02.016

Sheridan RP, Wang WM, Liaw A, Ma J, Gifford EM (2016) Extreme gradient boosting as a method for quantitative structure-activity relationships. J Chem Inf Model 56(12):2353–2360. https://doi.org/10.1021/acs.jcim.6b00591

Sillers WS, Fredlund DG (2001) Statistical assessment of soil-water characteristic curve models for geotechnical engineering. Can Geotech J 38:1297–1313. https://doi.org/10.1139/cgj-38-6-1297

Silvestrini RT, Montgomery DC, Jones B (2013) Comparing computer experiments for the Gaussian process model using integrated prediction variance. Qual Eng 25(2):164–174. https://doi.org/10.1080/08982112.2012.758284

Srivastava A, Babu GLS, Haldar S (2010) Influence of spatial variability of permeability property on steady state seepage flow and slope stability analysis. Eng Geol 110(3–4):93–101. https://doi.org/10.1016/j.enggeo.2009.11.006

van Genuchten MT (1980) A closed-form equation for predicting the hydraulic conductivity of unsaturated soils. Soil Sci Soc Am J 44:892–898. https://doi.org/10.2136/sssaj1980.03615995004400050002x

Vanapalli SK, Fredlund DG, Pufahl DE, Clifton AW (1996) Model for the prediction of shear strength with respect to soil suction. Can Geotech J 33(3):379–392. https://doi.org/10.1139/t96-060

Vanmarcke EH (1983) Random fields: analysis and synthesis. MIT Press, Cambridge

Wang B, Chen YL, Wu C, Peng Y, Song JJ, Liu WJ, Liu X (2018a) Empirical and semi-analytical models for predicting peak outflows caused by embankment dam failures. J Hydrol 562:692–702. https://doi.org/10.1016/j.jhydrol.2018.05.049

Wang L, Cao ZJ, Li DQ, Phoon KK, Au SK (2018b) Determination of site-specific soil-water characteristic curve from a limited number of test data—a Bayesian perspective. Geosci Front 9(6):1665–1677. https://doi.org/10.1016/j.gsf.2017.10.014

Wang L, Zhang WG, Chen FY (2019a) Bayesian approach for predicting soil-water characteristic curve from particle-size distribution data. Energies 12:2992. https://doi.org/10.3390/en12152992

Wang Y, Cao ZJ, Li DQ (2016) Bayesian perspective on geotechnical variability and site characterization. Eng Geol 203:117–125. https://doi.org/10.1016/j.enggeo.2015.08.017

Wang Y, Cao ZJ, Au SK (2010) Efficient Monte Carlo Simulation of parameter sensitivity in probabilistic slope stability analysis. Comput Geotech 37(7–8):1015–1022. https://doi.org/10.1016/j.compgeo.2010.08.010

Wang Y, Cao ZJ, Au SK (2011) Practical reliability analysis of slope stability by advanced Monte Carlo simulations in a spreadsheet. Can Geotech J 48(1):162–172. https://doi.org/10.1139/T10-044

Wang Y, Zhao TY, Phoon KK (2018c) Direct simulation of random field samples from sparsely measured geotechnical data with consideration of uncertainty in interpretation. Can Geotech J 55(6):862–880. https://doi.org/10.1139/cgj-2017-0254

Wang Y, Zhao TY, Hu Y, Phoon KK (2019b) Simulation of random fields with trend from sparse measurements without detrending. J Eng Mech 145(2):1–12. https://doi.org/10.1061/(ASCE)EM.1943-7889.0001560

Wong TT (2015) Performance evaluation of classification algorithms by k-fold and leave-one-out cross validation. Pattern Recognit 48(9):2839–2846. https://doi.org/10.1016/j.patcog.2015.03.009

Zhang WG, Goh ATC (2013) Multivariate adaptive regression splines for analysis of geotechnical engineering systems. Comput Geotech 48:82–95. https://doi.org/10.1016/j.compgeo.2012.09.016

Zhang WG, Goh ATC (2016) Multivariate adaptive regression splines and neural network models for prediction of pile drivability. Geosci Front 7(1):45–52. https://doi.org/10.1016/j.gsf.2014.10.003

Zhao LH, Zuo S, Lin YL, Li L, Zhang YB (2016) Reliability back analysis of shear strength parameters of landslide with three-dimensional upper bound limit analysis theory. Landslides 13(4):711–724. https://doi.org/10.1007/s10346-015-0604-3

Chapter 10
Efficient Time-Variant Reliability Analysis of Bazimen Landslide in the TGRA Using XGBoost and LightGBM

10.1 Introduction

Landslides are one of the most frequent geological disasters in China, which may cause considerably detrimental social and economic impact and have attracted increasing concern in geotechnical engineering practice (e.g., Peng et al. 2015; Huang et al. 2018; Wang et al. 2019; Ji et al. 2020). As one of the key areas for the prevention of landslide geological disasters in China, there are more than 4000 landslides distributed in the TGRA of Yangtze River (e.g., Yin et al. 2010; Tang et al. 2019; Guo et al. 2020), which poses great threat to the safety of local residents, ecological environments, and geotechnical engineering infrastructures. Rational stability analysis of reservoir slopes is a significant prerequisite for landslide prevention and mitigation in the TGRA. Although the deterministic slope stability analysis can be conveniently performed to measure the safety margin of slope stability using the index of *FS*, there is a growing consensus that there exist various uncertainties in geotechnical engineering (e.g., the inherent variability of geomaterial properties, transformation uncertainty, and measurement error), and the commonly used deterministic slope stability analysis methods are unable to take the geotechnical uncertainties into account (e.g., Phoon and Kulhawy 1999; Cao et al. 2016; Wang et al. 2016). In such a case, geotechnical reliability analysis opens up a novel way to consider the underlying various geotechnical uncertainties explicitly and to further quantify the safety margin of geotechnical structures by reliability index (or equivalently, failure probability) from a probabilistic perspective (e.g., Griffiths and Fenton 2004; Li et al. 2016a; Jiang et al. 2020).

In the past few decades, many researchers have contributed to the reliability analysis of geotechnical structures (e.g., slopes, tunneling, and deep excavations), and the number of relevant publications has been increasing at an accelerating pace (e.g., Li et al. 2016a, b; Liu and Cheng 2016; Ji et al. 2018; Liu et al. 2018, 2019; Yuan et al. 2019; He et al. 2020; Wang et al. 2020a, b; Liao and Ji 2021; Wang and Goh 2021). It is well recognized that periodic reservoir water level fluctuation and seasonal rainfall are the two main influential factors that significantly affect the stability of reservoir

Z. Wengang et al., *Application of Machine Learning in Slope Stability Assessment*,
https://doi.org/10.1007/978-981-99-2756-2_10

slopes (e.g., Song et al. 2018; Li et al. 2019), which may trigger the reactivation of ancient landslides or even new landslides in the TGRA (e.g., Huang et al. 2018; Tang et al. 2019; Zhang et al. 2020). Thus, it is of great practical significance to perform the reliability analysis of the reservoir slope stability considering the combined effects of the reservoir water level fluctuation and seasonal rainfall. Due to the variation property of the reservoir water level and rainfall, the reservoir slope reliability may be varying with the external environment. Under such circumstances, performing time-variant reliability of reservoir slope stability may be more suitable. Compared with the time-invariant reliability analysis (or static reliability analysis) that focuses on the evaluation of reliability index or failure probability at a specified time instant of interest, the time-variant reliability analysis (or equivalently, dynamic reliability analysis) generally pays attention to the temporal evolution of reliability index or failure probability during a given time period (e.g., Andrieu-Renaud et al. 2004; Cheng and Lu 2019; Ling et al. 2019; Du et al. 2019).

Until now, most of the previous studies were concentrated on the commonly used time-invariant reliability analysis. In contrast, little attention has been paid to the time-variant reliability analysis of reservoir slope stability. Wu et al. (2017) and Liao et al. (2021) contributed to perform the time-variant reliability analyzes of Huangtupo No. 2 landslide and Majiagou landslide considering the reservoir water level fluctuation through a specially designed reservoir water head. Both of them focused more on the influence of reservoir water level fluctuation on the landslide failure probability, and the effect of seasonal rainfall was not taken into account. Moreover, the MCS was used in their studies for evaluating the time-variant failure probability. Although the crude MCS has the merits of simplicity and flexibility in the evaluation of failure probability, it necessitates a large number of simulations to achieve an estimated reliability index or failure probability with satisfactory accuracy, leading to the known criticism of poor computational efficiency (e.g., Li et al. 2016a; Huang et al. 2017a, b). Generally, the time-variant reliability analysis can be transformed into a series of time-invariant reliability analyzes at discrete time instants (e.g., Ling et al. 2019; Straub et al. 2020). This means that the computational complexities of the time-variant reliability analysis will be proportional to both the number of discrete time instants and the associated computational efforts at each time instant. In other words, if extending the crude MCS to the time-variant reliability analysis directly, the considerable computational efforts may be a prohibitively expensive task in practical applications. This calls for an efficient approach to facilitate the time-variant reliability analysis of reservoir slope stability.

With the rapid development of artificial intelligence technologies, more and more ML algorithms and their variants have been successfully applied in the geotechnical reliability analysis, such as extreme learning machine (e.g., Kumar and Samui 2019; Ling et al. 2021), MARS (e.g., Liu et al. 2019; Wang et al. 2020a; Deng et al. 2021), XGBoost (e.g., Wang et al. 2020b, c), and CNN (Wang et al. 2021; Wang and Goh 2021). These ML algorithms greatly improve the computational efficiency in the geotechnical reliability analysis, allowing geotechnical engineers and researchers to focus more on the engineering problems without being comprised by the prohibitively computational tasks. To the best of our knowledge, although the ML has many successful applications in the geotechnical reliability analysis, it is rarely applied in the time-variant reliability analysis of geotechnical structures. The integration of ML may offer a promising approach to promote the time-variant reliability analysis of reservoir slope stability.

This study aims to propose an efficient time-variant reliability analysis approach for facilitating the evaluation of Bazimen landslide time-variant failure probability based on the advanced machine learning algorithms of XGBoost and LightGBM. The remainder of this part starts with the introduction of transient seepage analysis and slope stability analysis. Subsequently, the XGBoost-based and LightGBM-based time-variant reliability analysis approaches are proposed, and their implementation procedures are outlined. Finally, the proposed approaches are applied to the time-variant reliability analysis of Bazimen landslide in the TGRA. The performances of them in the evaluation of Bazimen landslide time-variant failure probability are systematically investigated.

10.2 Methodology

10.2.1 Extreme Gradient Boosting

XGBoost is an advanced supervised algorithm proposed by Chen and Guestrin (2016) under the tree boosting framework, which constructs a tree ensemble model consisting of several regression trees to approximate the input–output mapping of interest. It introduces the additive learning process to build all the regression trees sequentially, and each newly added tree learns from its former trees and updates the residuals in the prediction values. Due to the advantages of high efficiency and sufficient flexibility, XGBoost has gained increasing popularity in the famous Kaggle machine learning competitions.

The target value $\widehat{y}_i$ predicted from the XGBoost is given by (Chen and Guestrin 2016):

$$\hat{y}_i = \sum_{k=1}^{K} f_k(\widehat{\Theta}_i), \quad f_k \in \mathcal{F} \tag{10.1}$$

where $\mathcal{F}$ denotes the space of regression trees; K is the total number of regression trees; $\widehat{\Theta}_i$ represents the features considered in machine learning; $f_k(\cdot)$ is the predicted value corresponding to the k-th tree. It can be observed that $\hat{y}_i$ is essentially the summation of the predicted values of all the regression trees. The total number of regression trees and the optimal structure of each tree can be determined by optimizing the predefined objective function Γ_{obj}. The Γ_{obj} can be defined as

$$\Gamma_{obj} = \sum_{i=1}^{n} l(y_i, \hat{y}_i) + \sum_{k=1}^{K} \Omega(f_k) \tag{10.2}$$

where n is the number of training data; $l(y_i, \hat{y}_i)$ is the loss function describing that how well the model fits the training data; $\Omega(f_k)$ is an additional regularization term for penalizing complicated models and avoiding the overfitting issue. It is generally infeasible to optimize the Γ_{obj} using the traditional optimization methods because it involves a tree ensemble model. As an alternative, the additive learning strategy is used in the XGBoost. The objective function Γ^t_{obj} for the t-th iteration is expressed as

$$\begin{aligned} \Gamma^t_{\text{obj}} &= \sum_{i=1}^{n} l(y_i, \hat{y}_i^t) + \sum_{k=1}^{t} \Omega(f_k) \\ &= \sum_{i=1}^{n} l(y_i, \hat{y}_i^{t-1} + f_t(\widehat{\Theta}_i)) + \Omega(f_t) + C_0 \end{aligned} \tag{10.3}$$

where C_0 represents a constant. By performing the second-order Taylor expansion, the Γ^t_{obj} can be approximated by

$$\Gamma^t_{\text{obj}} \approx \sum_{i=1}^{n} \left[l(y_i, \hat{y}_i^{t-1}) + g_i f_t(\widehat{\Theta}_i) + \frac{1}{2} h_i f_t^2(\widehat{\Theta}_i) \right] + \Omega(f_t) + C_0 \tag{10.4}$$

where $g_i = \partial l(y_i, \hat{y}_i^{t-1})/\partial \hat{y}_i^{t-1}$ and $h_i = \partial^2 l(y_i, \hat{y}_i^{t-1})/\partial (\hat{y}_i^{t-1})^2$ denote the first and second-order partial derivatives of the loss function, respectively. After removing the constants, the Γ^t_{obj} can be further reformulated as

$$\Gamma^t_{obj} = \sum_{i=1}^{n} \left[g_i f_t(\widehat{\Theta}_i) + \frac{1}{2} h_i f_t^2(\widehat{\Theta}_i) \right] + \Omega(f_t) \tag{10.5}$$

The optimal model parameters of the t-th tree can be determined by optimizing the objective function Γ^t_{obj} (i.e., Eq. (10.5)), and then the corresponding predicted value is also available. Determination of the split point is a prerequisite in tree learning, and the XGBoost uses the exact greedy algorithm and approximate algorithm to search the best split point among the possible split points. More detailed explanations of the XGBoost algorithm can refer to Chen and Guestrin (2016).

10.2.2 Light Gradient Boosting Machine

LightGBM is a novel variant of the GBDT developed by Microsoft in 2017 (Ke et al. 2017) for tackling the challenges in learning the decision trees when the data size is large, and the feature dimension is high, which has gained increasing attention in many areas (e.g., Fan et al. 2019; Fang et al. 2021; Massaoudi et al. 2021) due to its excellent performance in the aspect of computational efficiency and memory consumption. Generally, the conventional GBDT models need to scan all the data to evaluate the information gain of all the possible split points for each feature, and hence, the computational efforts may become prohibitively expensive for the problems involving big data and numerous features. In contrast, LightGBM introduces two advanced techniques called GOSS and EFB to address this issue by reducing the number the data instances and features rationally.

GOSS reduces the number of data instances by discarding the proportion of data instances with small gradients and only use the rest to evaluate the information gain. The reason is that the gradient of data instance poses a significant influence on the evaluation of information gain, and the data instances with small gradients will provide marginal contributions in the estimation of information gain. In the implementation of GOSS, the data instances are firstly ranked in descending order according to the absolute values of gradients, and the top $a \times 100\%$ data instances are assigned to a sub-set A. Then, build a new sub-set B with $b \times |A^c|$ from the complementary set A^c through random sampling. Finally, the variance gain $\tilde{V}_j(d)$ of feature j at the point d can be used to split the data instances, which is written as (Ke et al. 2017):

$$\tilde{V}_j(d) = \frac{1}{n}\left(\frac{\left(\sum_{x_i \in A_l} g_i + \frac{1-a}{b}\sum_{x_i \in B_l} g_i\right)^2}{n_l^j(d)} + \frac{\left(\sum_{x_i \in A_r} g_i + \frac{1-a}{b}\sum_{x_i \in B_r} g_i\right)^2}{n_r^j(d)}\right) \tag{10.6}$$

where $n_l^j(d)$ and $n_r^j(d)$ denote the number of nodes on the left and right, respectively; $(1-a)/b$ is a coefficient used to normalize the gradients; A_l, A_r, B_l, and B_r are the sub-sets defined based on the sub-set A and B.

The basic idea of EFB is to reduce the number of features by bundling mutually exclusive features. Generally, many features may be mutually exclusive in a sparse feature space, and these mutually exclusive features are unable to take nonzero values simultaneously. Thus, the mutually exclusive features can be rationally bundled into a single feature, enabling the reduction of features and the enhancement of computational efficiency. To strike a balance between computational accuracy and efficiency, the conflict rate is frequently used to assess whether to bundle the features together or not (Fang et al. 2021). More detailed explanations of the LightGBM algorithm can refer to Ke et al. (2017).

10.2.3 Hyperparameter Optimization

Determination of the hyperparameters is a primary concern in machine learning applications, which has a significant influence on predictive performance (Zhang et al. 2021). The grid search (GS) and random search (RS) methods are frequently used to tune the hyperparameters. However, the GS suffers a known deficiency of poor efficiency, because its computational burden will increase exponentially with the number of hyperparameters in the machine learning algorithm. Thus, the GS may be infeasible to the XGBoost and LightGBM since both of them have many hyperparameters. Different from the GS that tries all the possible combinations of the hyperparameters, RS accelerates the search of optimal hyperparameters by trying the random combinations. However, the main drawback of the RS is that it may miss the optimal values (Massaoudi et al. 2021). In contrast, Bayesian optimization is a preferable choice due to its efficiency and scalability in tackling hyperparameter optimization issues. Specifically, the sequential model-based optimization method is used in this study, which is able to provide the global optimal hyperparameters with the minimum loss. For the sake of brevity, details of the sequential model-based optimization are not provided here, which can be referred to Zhang et al. (2021).

10.2.4 Evaluation Indicators

To quantitatively measure the predictive performance of the established ML models, several commonly used evaluation indicators are adopted in this study, including RMSE and R^2. Accordingly, their mathematical expressions are given below (e.g., Zhang et al. 2021; Zhou et al. 2021):

$$\text{RMSE} = \sqrt{\frac{1}{N}\sum_{i=1}^{N}\left(y_i - \hat{y}_i\right)^2} \tag{10.7}$$

$$R^2 = 1 - \frac{\sum_{i=1}^{N}\left(y_i - \hat{y}_i\right)^2}{\sum_{i=1}^{N}\left(y_i - \overline{y}\right)^2} \tag{10.8}$$

where N is the total number of data, y_i and $\hat{y}_i$ denote the observed values and predicted values, respectively; $\overline{y}$ represents the mean of the observed values. As indicated by Eqs. (10.7) and (10.8), the closer the RMSE value is to 0 indicating that the more accurate the predictive model is. The larger the value of R^2, the better the predictive model.

10.3 ML-Based Time-Variant Reliability Analysis

10.3.1 Monte Carlo Simulation

The failure probability (or equivalently, reliability index) is an index used to quantify the safety margin of geotechnical structures probabilistically, and the evaluation of it is a primary concern in the geotechnical reliability analysis. Since the complicated implicit performance functions are usually encountered in geotechnical engineering practice, probably leading to the unavailability of analytical solutions. In such a case, an approximate numerical calculation method called MCS has gained popularity in evaluating the failure probability by virtue of its conceptual simplicity and easy to use for geotechnical researchers and engineers.

Consider, for example, the soil parameters are regarded as random variables in this study for characterizing the underlying geotechnical parameter uncertainty. Generally, the random variables of soil parameters are assumed to follow a probability density function (PDF) (e.g., Li et al. 2015), such as normal distribution and lognormal distribution. Based on the predefined PDF, a total number of N_{MCS} random variables $\widehat{X}$ can be generated. Then, the factor of safety (FS) corresponding to each set of random variable sample can be calculated by repeatedly invoking commercial geotechnical software to perform deterministic slope stability analysis. Finally, the failure probability P_f can be calculated as

$$P_f = \frac{1}{N_{\text{MCS}}} \sum_{i=1}^{N_{\text{MCS}}} I\left[FS(\widehat{X}_i) < 1\right] \tag{10.9}$$

where $\text{FS}(\widehat{X}_i)$ is the FS value corresponding to i-th set of random variable sample $\widehat{X}_i$; $I[\cdot]$ is an indicator function for judging whether slope failure occurs. If $\text{FS}(\widehat{X}_i) < 1$, $I[\cdot] = 1$; otherwise, $I[\cdot] = 0$. The COV of P_f (i.e., $\text{COV}_{P_f} = \sqrt{\left(1 - P_f/\left(N_{\text{MCS}} \cdot P_f\right)\right)}$) is usually regarded as a guideline for determining the number of simulations (i.e., N_{MCS}) (e.g., Li et al. 2016a; Wang and Goh 2021).

As indicated by Eq. (10.9), the FS values evaluated from the deterministic slope stability analysis provide a basis for the evaluation of failure probability P_f. It is well recognized that the stability of reservoir slopes is significantly affected by periodic reservoir water level fluctuation and seasonal rainfall in the TGRA. Therefore, accurate analysis of seepage flow through the reservoir slopes is of great importance for the stability evaluation, which provides valuable seepage information (e.g., pore water pressure) for the subsequent slope stability analysis. Until now, several commercial geotechnical software (e.g., Geostudio and Abaqus) have embedded function modules for performing deterministic seepage and slope stability analysis. Taking the Geostudio software (GEO-SLOPE International Ltd. 2012), for example, the SEEP/W module can be used to analyze the transient seepage flow through the slopes, and the SLOPE/W module is then applied to conduct slope stability analyzes by limit equilibrium methods (e.g., Morgenstern–Price) for calculating the *FS*.

Generally, the MCS method requires a large number of simulations to achieve the desired accuracy of the estimated P_f (e.g., Li et al. 2016a; Huang et al. 2017a, b). Although existing commercial geotechnical software offers a powerful tool for conducting seepage and slope stability analysis, it is a cumbersome task to repeatedly perform a large number of deterministic analyzes in practice applications when using the MCS due to its poor computational efficiency. To address this issue, the XGBoost-based and LightGBM-based reliability analysis approach are developed to facilitate the evaluation of failure probability, as described in the next subsection.

10.3.2 Calculation of Time-Variant Failure Probability

The main idea of ML in aiding geotechnical reliability analysis is to reconstruct the high-dimensional implicit performance function through learning from the prepared database which generally includes the input random variables or random field samples of geomaterial properties (e.g., shear strength parameters and hydraulic parameters) as well as the output quantity of interest (e.g., the *FS* in this study) that usually calculated from the commercial geotechnical software. As the ML-based reliability analysis model reaches the desired performance after sufficient training and rational validation, it can be readily used to evaluate the failure probability of geotechnical structures with reasonable accuracy and efficiency.

Inspired by the previous studies (e.g., Ling et al. 2019; Straub et al. 2020), the time-variant reliability analysis can be transformed into a series of time-invariant reliability analyzes at discrete time instants. In other words, the time-variant reliability analysis of geotechnical structures can be performed by conducting many time-invariant reliability analyzes (i.e., the commonly encountered static reliability analysis) at the discrete time instants during a given time period. At each discrete time instant, the MCS method described above can be applied to the time-invariant reliability analysis for calculating the failure probability (i.e., Eq. 10.9). Thereafter, the failure probability values corresponding to all the discrete time instants constitute the results analyzed in the time-variant reliability analysis.

In this study, an initial XGBoost model is firstly calibrated based on a preparatory database which contains the boundary conditions (i.e., reservoir water level and rainfall intensity), random variables of soil parameters (i.e., effective cohesion, effective friction angle, and saturated hydraulic conductivity) as well as the *FS* values calculated from the commercial geotechnical software. Then, the established XGBoost model is heuristically tuned for searching the optimal hyperparameters with the aid of Bayesian optimization technique, so as to achieve a desired performance. Finally, the *FS* values and corresponding failure probabilities at other time instants during a specific time period of interest can be readily evaluated using the established XGBoost model. Similarly, a LightGBM model can also be constructed and then applied to predict the time-variant failure probabilities of reservoir slopes during other time periods.

The integration of ML and MCS in the time-variant reliability analysis not only inherits the merits of MCS (i.e., flexibility and robust) but also can make the best use of the powerful learning ability of ML algorithms to portray the high-dimensional implicit performance function that is frequently encountered in geotechnical engineering practice, which alleviates the considerable computational burden underlying the evaluation of failure probabilities for all the discrete time instants. The detailed implementation procedures involved in the XGBoost-based and LightGBM-based time-variant reliability analysis approach are outlined in the next section.

10.4 Implementation Procedure

To facilitate the understanding and application of the proposed XGBoost-based and LightGBM-based time-variant reliability analysis approach in engineering practice for geotechnical engineers, Fig. 10.1 portrays the flowchart of them. Firstly, some necessary information needs to be provided, such as boundary conditions (e.g., reservoir water level and rainfall intensity) and statistics of soil parameters (e.g., PDFs, means, and standard deviations). Then, generate a total of N_{MCS} random variable samples using the MCS method, and the results of FS corresponding to these random variable samples at each time instant during the given time period T_a can be evaluated through existing commercial geotechnical software. In this study, the GeoStudio software is applied to perform transient seepage analysis and slope stability analysis at each time instant. Thereafter, these FS values associated with the inputs (i.e., random variable samples and boundary conditions) are regarded as training datasets for calibrating the XGBoost model and LightGBM model. With the aid of the constructed XGBoost model and LightGBM model, the FS values corresponding to the random variable samples at the time instants during other time periods of interest (e.g., T_b) can be conveniently obtained, and the time-variant failure probability at each time instant can also be calculated using Eq. (10.9). For illustration, the proposed approach is applied to the time-variant reliability analysis of a practical case adapted from the Bazimen landslide in the TGRA.

10.5 Application to Bazimen Landslide in the TGRA

The Bazimen landslide is located in Zigui County of Hubei Province, China, and on the right side of Xiangxi River which is a major tributary of the Yangtze River. The elevation of the landslide ranges from 110 to 250 m, and most of the landslides are submerged. The maximum longitudinal dimension is about 380 m, and the width varies from 100 to 350 m, giving rise to an estimated landslide volume of about 2 million cubic meters. The Bazimen landslide is mainly constituted by loose

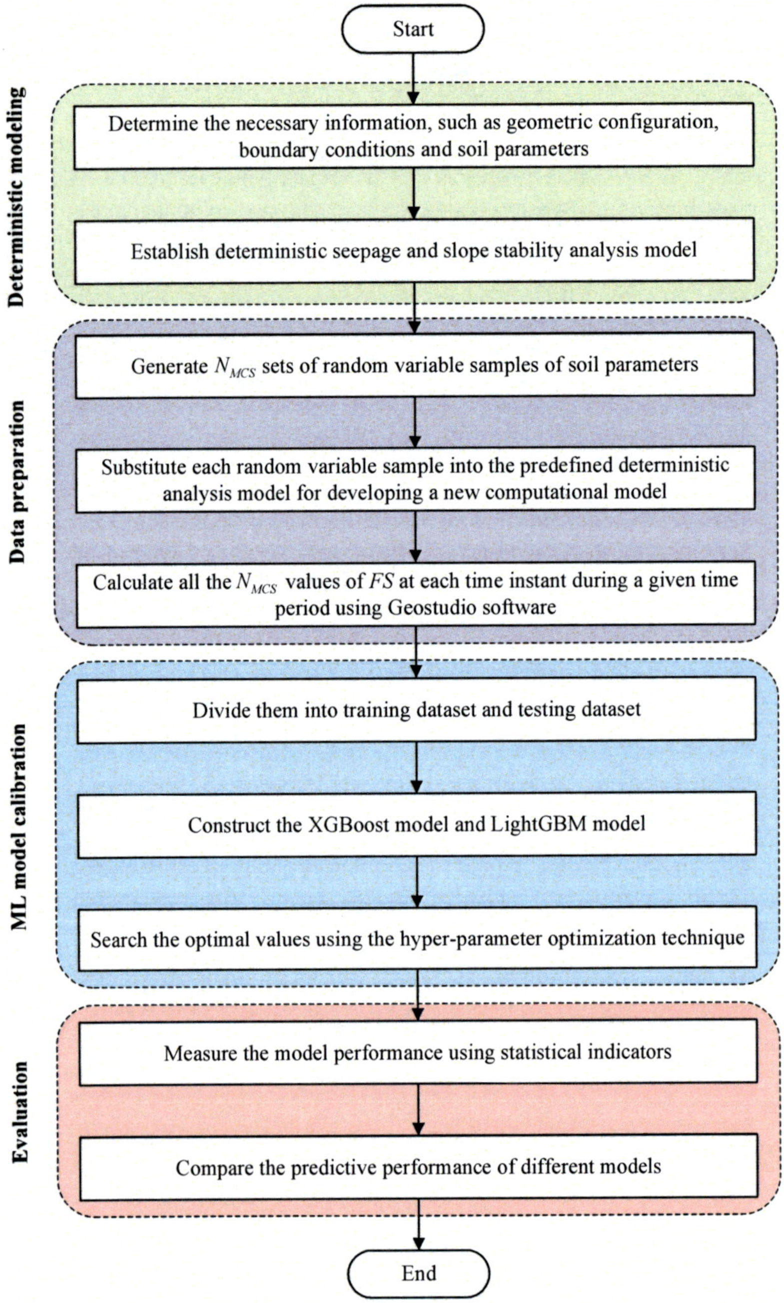

Fig. 10.1 Flowchart for the machine learning-based time-variant reliability analysis approach

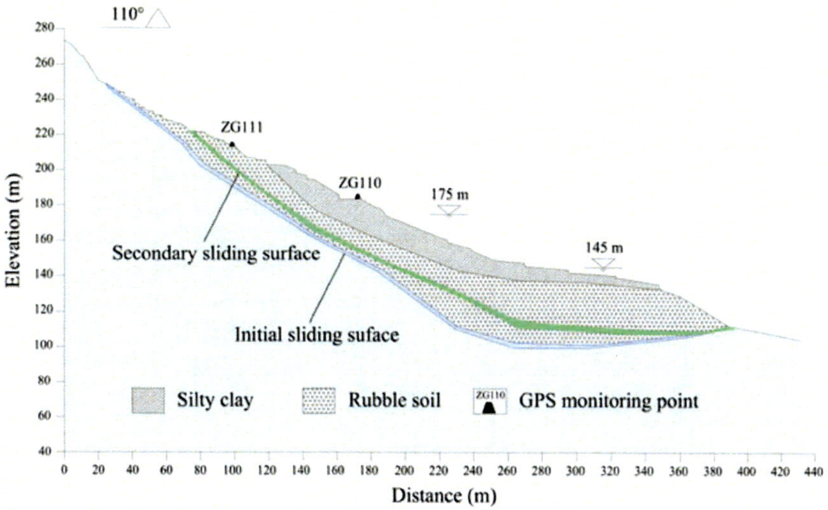

Fig. 10.2 Schematic geological cross-section of the Bazimen landslide

Quaternary deposits and fragmented rubble, and the bedrock is formed by clastic sediments with siltstone and sandstone. Figure 10.2 shows a schematic geological cross-section of the Bazimen landslide, and there exist two main sliding surfaces in the Bazimen landslide, which are denoted as the initial sliding surface and secondary sliding surface, respectively. The initial sliding surface is the lower one lying on the interface between soil deposits and bedrock. Inspire by the previous studies (e.g., Du et al. 2013; Huang et al. 2017a, b; Yang et al. 2019), the deformation behavior of the Bazimen landslide is significantly influenced by the seasonal rainfall and reservoir water level fluctuation. Thus, it is necessary to perform the stability analysis of Bazimen landslide considering both the effects of reservoir water level fluctuation and seasonal rainfall.

The seepage through the Bazimen landslide can be analyzed by performing transient seepage analysis. A varying total head equal to the reservoir water level is applied to the slope surface below the water level, which will fluctuate between 145 and 175 m. A flux boundary equal to the recorded rainfall intensity is applied to the slope surface above the water level. By assuming that the Bazimen landslide is underlain by an impermeable bedrock, a zero flux condition is applied to the bottom boundary. Figure 10.3 plots the historical records of rainfall and reservoir water level during the period 2008–2018 in the Bazimen landslide, which provide the necessary information for defining the hydraulic conditions in the transient seepage analysis.

Table 10.1 shows the soil parameters of each layer used in this study, which are adapted from the previous research on similar projects in the TGRA. To consider the parameter uncertainties in the time-variant reliability analysis of Bazimen landslide, the three main soil parameters (i.e., the saturated hydraulic conductivity, effective cohesion, and friction angle) are regarded as random variables. The COVs, known

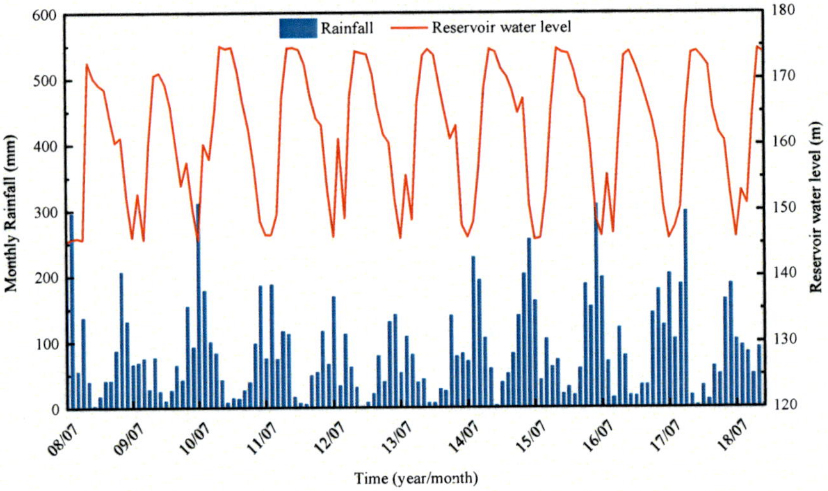

Fig. 10.3 Rainfall and reservoir water level during the period 2008–2018 in the Bazimen landslide

as the ratio between standard deviation and mean, and the PDFs of soil parameters required in the random variable characterization are usually unavailable in engineering practice due to the limited site-specific test data. Thus, the COVs and PDFs of the saturated hydraulic conductivity, effective cohesion, and friction angle are collected from the literature and assumed to have the same value for different soil layers in this study (e.g., Cho 2012; Li et al. 2015). Among several parametric soil–water characteristic curve models, the van Genuchten (1980)-Mualem (1976) model (abbreviated as VGM) is used in this study for describing the hydraulic behavior of unsaturated soils.

In this study, the SEEP/W and SLOPE/W modules contained in the GeoStudio software are applied to conduct transient seepage analysis and slope stability analysis of the Bazimen landslide, respectively. Since the number of MCS (i.e., N_{MCS}) influences the evaluation of failure probability, it is necessary to choose an appropriate number to achieve an estimate with the desired accuracy. It is observed that the P_f varies slightly when the number of MCS is larger than 1000, and the corresponding COV is also less than 10%. Thus, $N_{\mathrm{MCS}} = 1000$ is adopted in this study to strike a balance between computational accuracy and efficiency. Based on the total number of 1000 random variable samples, the corresponding FS values can be evaluated by repeatedly invoking the GeoStudio software, and then, the time-variant failure probability of Bazimen landslide at each time instant is calculated using the Eq. (10.9). Figure 10.4 plots the time-variant failure probability of Bazimen landslide during the period 2008–2018. These results associated with the inputs can be used to compile a database for calibrating the XGBoost model and LightGBM model. In this study, the datasets ranging from July 2008 to December 2017 are regarded as training data, and the remaining datasets (i.e., ranging from January 2018 to December 2018) are considered as testing data.

Table 10.1 Statistical information of soil properties for Bazimen landslide

Parameter		Initial sliding surface	Secondary sliding surface	Rubble soil	Silty clay
k_s (m/s)	Mean	5.79×10^{-7}	6.48×10^{-7}	1.64×10^{-6}	2.27×10^{-6}
	COV	0.5			
	Distribution	Lognormal			
φ' (°)	Mean	19	19.7	31.8	26.2
	COV	0.2			
	Distribution	Lognormal			
c' (kPa)	Mean	17.6	18.3	42.1	32.4
	COV	0.3			
	Distribution	Lognormal			
γ_{uns} (kN/m^3)	–	20.6	20	21.1	20.1
γ_{sat} (kN/m^3)		21.5	20.9	22.3	20.7
α_{vm} (kPa)	–	100	100	20	20
n_{vm}		1.23	1.21	1.56	1.41
θ_s		0.35	0.33	0.37	0.43
θ_r		0.06	0.04	0.05	0.02

Note k_s is the saturated hydraulic conductivity; c' is the effective cohesion; φ' is the effective friction angle; γ_{uns} is the unsaturated unit weight; γ_{sat} is the saturated unit weight; α_{vm} and n_{vm} are the fitting parameters of VGM model; θ_s and θ_r are the saturated volumetric water content and residual volumetric water content, respectively; symbol "–" denotes the not applicable item

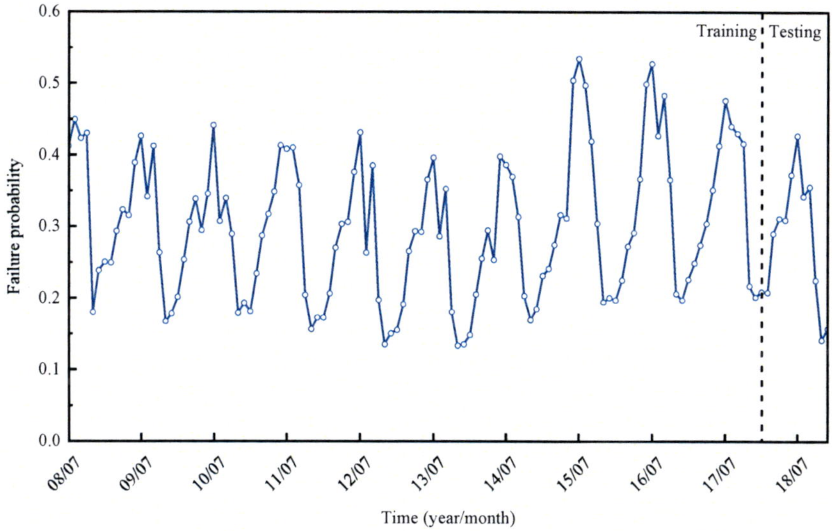

Fig. 10.4 Time-variant failure probability of Bazimen landslide during the period 2008–2018

10.5.1 Construction of XGBoost and LightGBM Models

Figure 10.5 plots the results of time-variant failure probability evaluated from the established XGBoost model and MCS for the training dataset and testing dataset. It is observed that the established XGBoost model can reasonably capture the data points calculated from the MCS method for both the training dataset and testing dataset. This reveals that the established XGBoost model is able to portray the variation of time-variant failure probability during the period 2008–2018. Furthermore, Fig. 10.6 shows the results of time-variant failure probability calculated from the established LightGBM model and MCS for the training dataset and testing dataset. The results obtained from the two methods are relatively closer, and the LightGBM model can rationally depict the changes of time-variant failure probability during the given time period, confirming that the LightGBM model performs well in the prediction of time-variant failure probability. Both the above XGBoost and LightGBM models are modeled and predicted on the Google Colaboratory platform, the former takes about 11.42 s, and the latter spends about 9.27 s. Although the database contains a total of 1.26 million datasets and 15 features in this Bazimen landslide example, these two models exhibit high efficiency in handling such a large number of data and high-dimensional features, which again validates their powerful capability. Thus, it can be concluded that both the XGBoost and LightGBM models are able to provide satisfactory performance in the prediction of time-variant failure probability for the Bazimen landslide, which offers a promising tool to facilitate the time-variant reliability analysis of reservoir slopes.

10.5.2 Performance of Model Averaging

Hybrid models have received increasing attention in machine learning, and several methods are frequently used to establish a hybrid model, such as averaging method, blending method, and stacking method. This subsection tries to investigate the performance of the model averaging method in the prediction of time-variant failure probability, where the above XGBoost and LightGBM models are used as basic models. Figure 10.7 plots the results of time-variant failure probability evaluated from the model averaging method and MCS for the training dataset and testing dataset. As expected, the data points obtained from the MCS method can be better captured by the averaging model for both the training dataset and testing dataset. This validates the effectiveness of the model averaging method in the prediction of Bazimen landslide time-variant failure probability, which suggests a possibility of promoting the time-variant reliability analysis by integrating several machine learning algorithms.

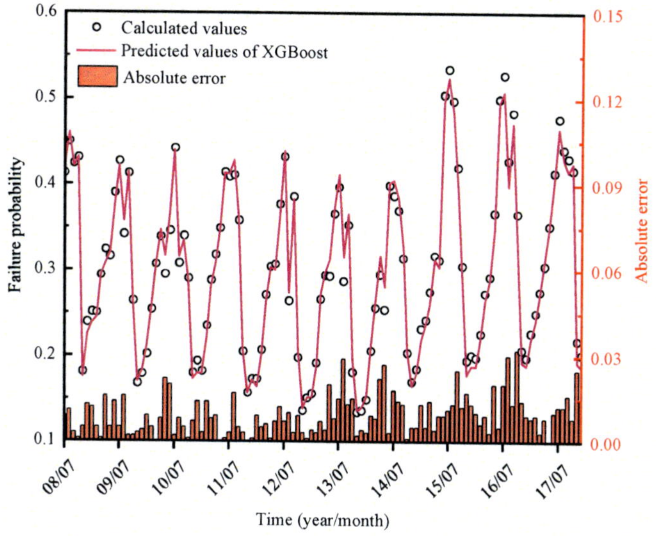

(a) Training dataset

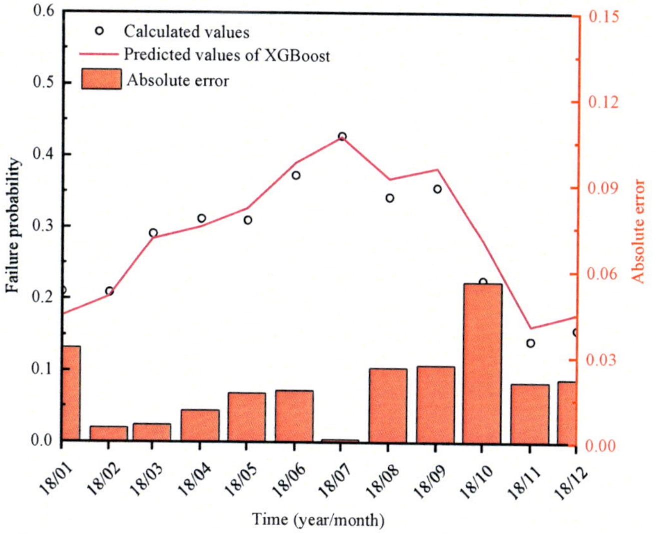

(b) Testing dataset

Fig. 10.5 Time-variant failure probability of Bazimen landslide predicted from the XGBoost model

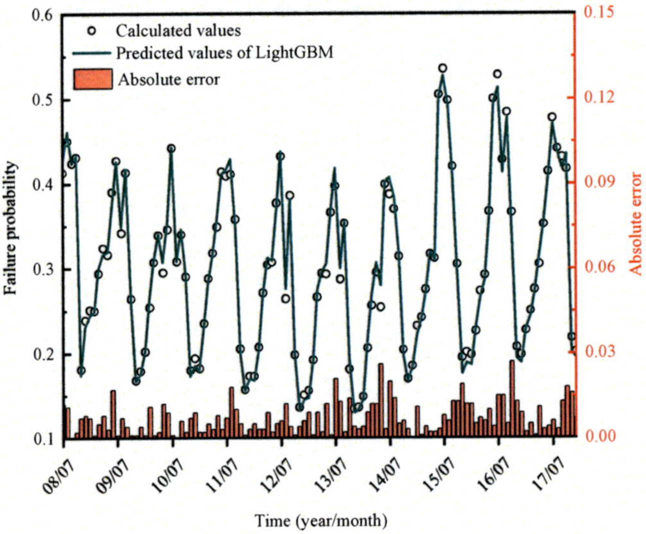

(a) Training dataset

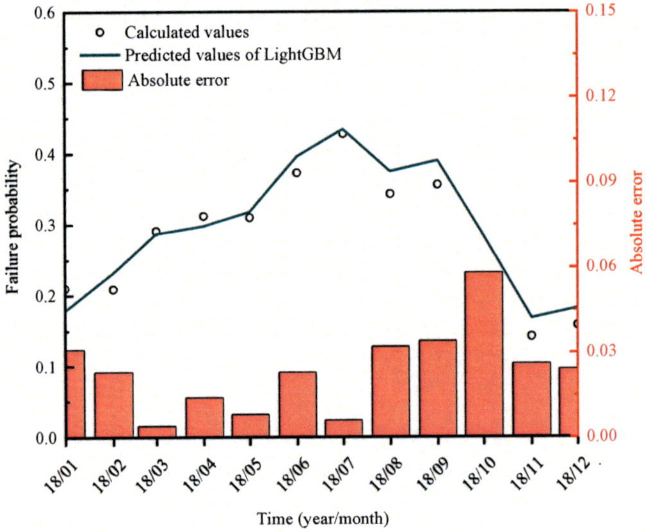

(b) Testing dataset

Fig. 10.6 Time-variant failure probability of Bazimen landslide predicted from the LightGBM model

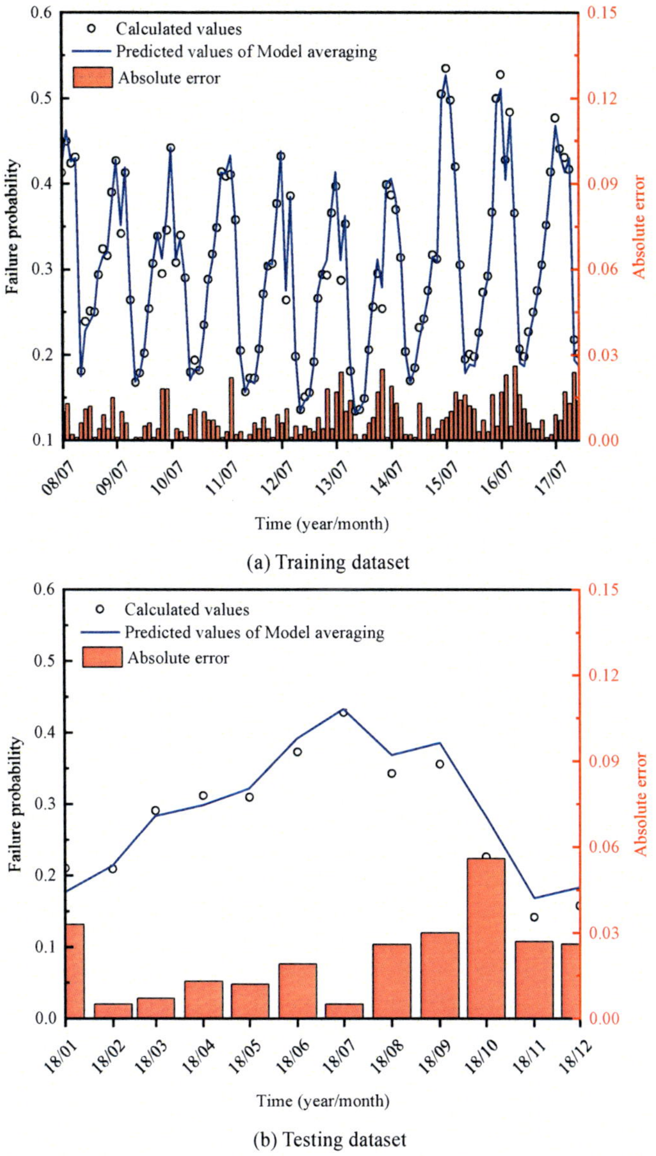

Fig. 10.7 Time-variant failure probability of Bazimen landslide predicted from the model averaging

10.5.3 Comparison of the Proposed Approaches and Prophet Model

The Prophet is a new time-series forecasting model developed by Facebook company for handling time-series data in the business community, which decomposes the time-series data into three components (i.e., trend, seasonality, and holidays) and then predicts it in an additive manner (Taylor and Letham 2018). In geotechnical and geological engineering, the landslide displacement prediction is a well-known time-series forecasting problem that has gained considerable attention in the past few decades (e.g., Du et al. 2013; Huang et al. 2017a, b; Yang et al. 2019). The prediction of Bazimen landslide time-variant failure probability can also be taken as a time-series forecasting issue, and this subsection attempts to explore the predictive performance of Prophet model and then to compare it with the above XGBoost model and LightGBM model.

Figure 10.8 plots the results of time-variant failure probability calculated from the Prophet model and MCS for the training dataset and testing dataset. In general, although the Prophet model can portray the changes of time-variant failure probability in the training dataset, it is unable to capture the data points well, and the associated absolute errors are also obviously larger than that calculated from the XGBoost model and LightGBM model. For the testing dataset, it is evident that there exists a great discrepancy between the results predicted from the Prophet model and the data points calculated from the MCS method. Besides, as shown in Fig. 10.8b, most of the absolute error values are larger than 0.05, which is also apparently greater than that of the XGBoost model and LightGBM model. Thus, the performance of Prophet model in the prediction of Bazimen landslide time-variant failure probability is unsatisfactory.

Furthermore, Fig. 10.9 compares the results of time-variant failure probability predicted from the XGBoost model, LightGBM model, and Prophet model. It can be observed that the XGBoost model and LightGBM model are able to reasonably portray the data points calculated from the MCS method for training dataset, and the RMSE values of them are 0.012 and 0.009, respectively. In contrast, the data points calculated from the MCS method cannot be well captured by the Prophet model, and the corresponding RMSE value is about 0.035, which is obviously greater than that of the XGBoost model and LightGBM model. For the testing dataset, compared with the XGBoost model and LightGBM model which performs well in the prediction of time-variant failure probability, the results of time-variant failure probability predicted from the Prophet models tend to be highly overestimated. Likewise, the Prophet model gets the maximum RMSE value among the three ML models (i.e., 0.091), and it is considerably larger than that of the other two models (i.e., 0.025 of the XGBoost model and 0.028 of the LightGBM model).

In this Bazimen landslide example, it can be concluded that the XGBoost model and LightGBM model perform relatively better than the Prophet model in the prediction of time-variant failure probability, which may be attributed to the different learning mechanisms underlying the three ML models. Both the XGBoost and

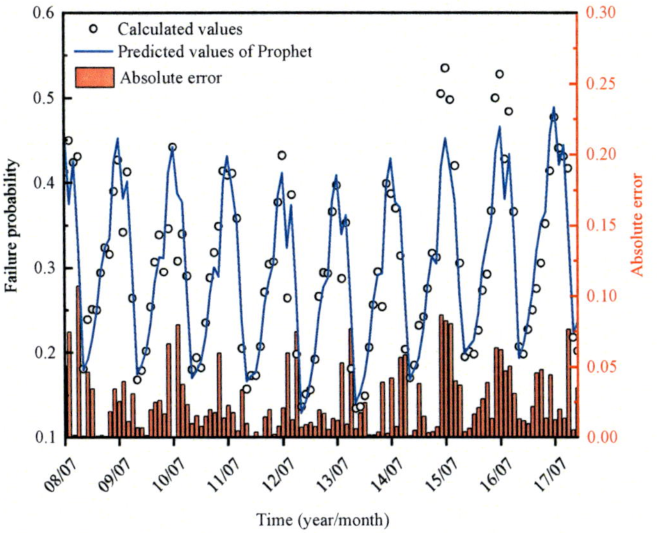

(a) Training dataset

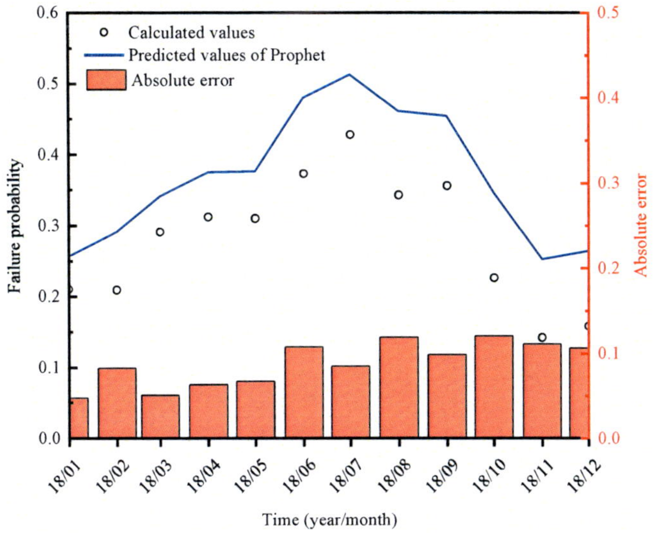

(b) Testing dataset

Fig. 10.8 Time-variant failure probability of Bazimen landslide predicted from the Prophet model

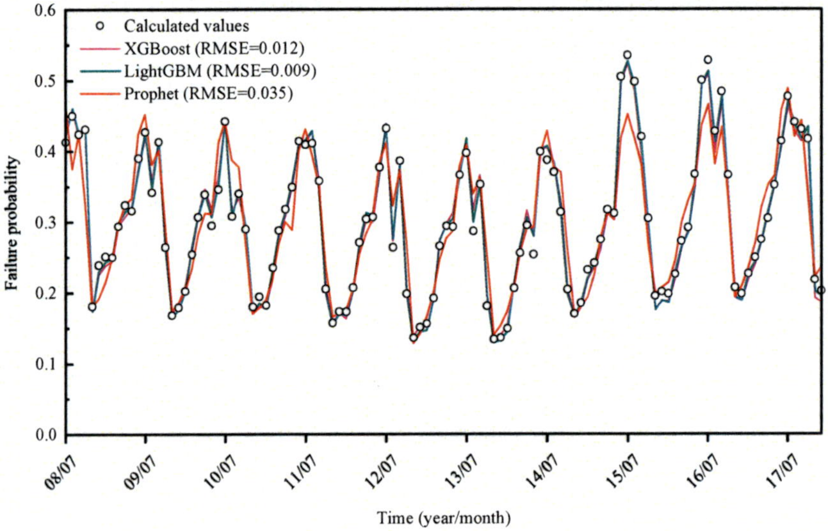

(a) Training dataset

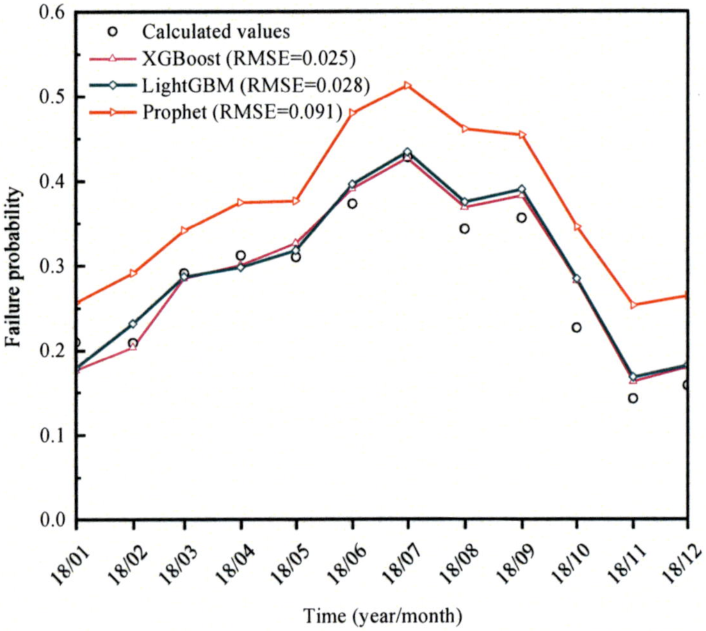

(b) Testing dataset

Fig. 10.9 Predictive performance of different machine learning models

LightGBM belong to gradient boosting algorithms, and they try to find the high-dimensional implicit relationships between the input variables (e.g., shear strength parameters) and the output (e.g., *FS*). With the aid of the well-established XGBoost and LightGBM model, the *FS* values and the time-variant failure probability at each time instant can be evaluated conveniently. In contrast, the Prophet model is a time-series model which makes use of the historical target information (i.e., time-variant failure probability at each time instant) to capture the underlying temporal evolution pattern, and then applied it to predict the time-variant failure probability at the next time instants. Thus, the knowledge of input geotechnical parameters is not a necessity in the Prophet model, which is the main difference between it and the other two models (i.e., XGBoost and LightGBM). The above results suggest the superiority of the XGBoost and LightGBM in predicting the time-variant failure probability of the Bazimen landslide example, and the Prophet model appears to be inferior to them in this study.

10.5.4 Feature Importance Analysis

The relative contribution of each feature to the model can be quantified by conducting feature importance analysis in machine learning. Generally, several indexes can be applied to measure the relative importance of features, such as 'weight', 'gain', and 'cover'. For illustration, the 'cover' criterion is used in this study. Figure 10.10 ranks the relative importance of all the fourteen input features based on the XGBoost model, and the arrangement of them from top to bottom is based on their relative importance. It can be seen that the φ' of the initial sliding surface obtains the maximum value of relative importance (i.e., 19.71%) among the fourteen features, followed by φ' of the rubble soil, reservoir water level, and rainfall. This means that the strength of the initial sliding surface plays a significant role in the evaluation of time-variant failure probability of the Bazimen landslide. Compared with geotechnical parameters, the reservoir water level and rainfall also pose a non-ignorable influence on the Bazimen landslide failure probability.

10.6 Summary and Conclusions

This chapter developed an efficient time-variant reliability analysis approach by introducing the advanced machine learning algorithms of XGBoost and LightGBM. With the aid of the constructed XGBoost and LightGBM model, the time-variant failure probability of landslides at the time instants of interest can be calculated accurately and efficiently, which avoids the prohibitive computational cost of repeatedly conducting a large number of deterministic analyzes at each time instant. The proposed approach was illustrated using a practical case adapted from the Bazimen landslide in the TGRA. The performances of XGBoost, LightGBM, and

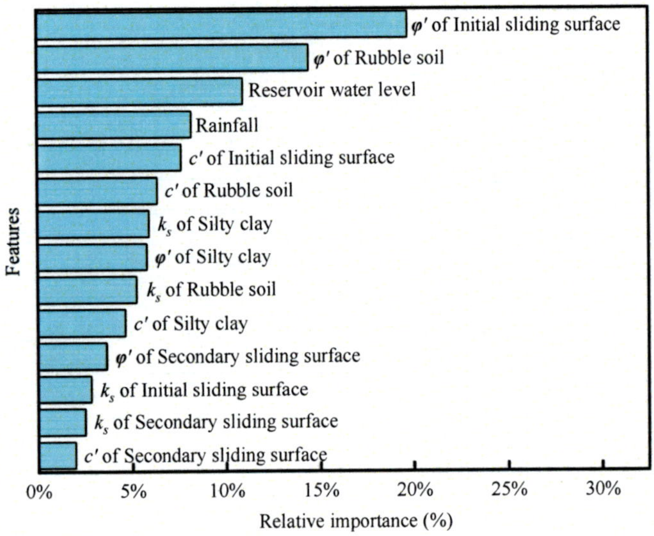

Fig. 10.10 Results of feature importance analysis

Prophet in the evaluation of Bazimen landslide time-variant failure probability were systematically investigated. The following conclusions can be drawn from this chapter:

1. The predicted FS values obtained from the established XGBoost and LightGBM model agree well with those calculated from the deterministic analyzes for both the training dataset and testing dataset. The proposed approach can rationally portray the changes of time-variant failure probability during the given time period, indicating that the XGBoost and LightGBM model are able to provide satisfactory performance in the prediction of time-variant failure probability for the Bazimen landslide, which offers a promising tool to facilitate the time-variant reliability analysis of reservoir slopes.
2. The averaging model inheriting the advantages of basic models can predict the Bazimen landslide time-variant failure probability with high accuracy. Compared with the Prophet model, the proposed approach performs relatively better than it in the prediction of time-variant failure probability, which may be attributed to the different learning mechanisms underlying them. Different from the XGBoost and LightGBM that try to find the high-dimensional input–output mapping relationship, the Prophet model is a time-series model which aims to capture the temporal evolution pattern of time-variant failure probability through making use of the historical information.
3. The results reveal the superiority of the proposed approach in predicting the time-variant failure probability of the Bazimen landslide. It offers an effective way to rationally evaluate the stability of reservoir slopes subjected to the combined

effects of rainfall and reservoir water fluctuation from a probabilistic perspective, which may provide preliminary guidance on landslide disaster prevention and mitigation in the TGRA. It is worth pointing out that the Bazimen landslide example was used in this study for illustration, the proposed approach can also be applied to other landslide cases of interest provided that the associated necessary information (e.g., geological conditions, geometric, and geotechnical parameters) is available.

References

Andrieu-Renaud C, Sudret B, Lemaire M (2004) The PHI2 method: a way to compute time-variant reliability. Reliab Eng Syst Saf 84(1):75–86. https://doi.org/10.1016/j.ress.2003.10.005

Cao ZJ, Wang Y, Li DQ (2016) Site-specific characterization of soil properties using multiple measurements from different test procedures at different locations—a Bayesian sequential updating approach. Eng Geol 211:150–161. https://doi.org/10.1016/j.enggeo.2016.06.021

Chen TQ, Guestrin C (2016) XGBoost: a scalable tree boosting system. In: Proceedings of the 22nd ACM SIGKDD international conference on knowledge discovery and data mining, 785–794. https://doi.org/10.1145/2939672.2939785

Cheng K, Lu Z (2019) Time-variant reliability analysis based on high dimensional model representation. Reliab Eng Syst Saf 188:310–319. https://doi.org/10.1016/j.ress.2019.03.041

Cho SE (2012) Probabilistic analysis of seepage that considers the spatial variability of permeability for an embankment on soil foundation. Eng Geol 133–134:30–39. https://doi.org/10.1016/j.enggeo.2012.02.013

Deng ZP, Pan M, Niu JT, Jiang SH, Qian WW (2021) Slope reliability analysis in spatially variable soils using sliced inverse regression-based multivariate adaptive regression spline. Bull Eng Geol Environ, 0123456789. https://doi.org/10.1007/s10064-021-02353-9

Du J, Yin K, Lacasse S (2013) Displacement prediction in colluvial landslides, Three Gorges Reservoir. China. Landslides 10(2):203–218. https://doi.org/10.1007/s10346-012-0326-8

Du W, Luo Y, Wang Y (2019) Time-variant reliability analysis using the parallel subset simulation. Reliab Eng Syst Saf 182:250–257. https://doi.org/10.1016/j.ress.2018.10.016

Fan J, Ma X, Wu L, Zhang F, Yu X, Zeng W (2019) Light Gradient boosting machine: an efficient soft computing model for estimating daily reference evapotranspiration with local and external meteorological data. Agric Water Manage 225:105758. https://doi.org/10.1016/j.agwat.2019.105758

Fang M, Chen Y, Xue R, Wang H, Chakraborty N, Su T, Dai Y (2021) A hybrid machine learning approach for hypertension risk prediction. Neural Comput Appl. https://doi.org/10.1007/s00521-021-06060-0

GEO-SLOPE International, Ltd. (2012) Geostudio. https://www.geoslope.com/

Griffiths DV, Fenton GA (2004) Probabilistic slope stability analysis by finite elements. J Geotech Geoenviron Eng 130:507–518. https://doi.org/10.1061/(ASCE)1090-0241(2004)130:5(507)

Guo Z, Chen L, Yin K, Shrestha DP, Zhang L (2020) Quantitative risk assessment of slow-moving landslides from the viewpoint of decision-making: a case study of the Three Gorges Reservoir in China. Eng Geol 273:105667. https://doi.org/10.1016/j.enggeo.2020.105667

He X, Xu H, Sabetamal H, Sheng D (2020) Machine learning aided stochastic reliability analysis of spatially variable slopes. Comput Geotech 126:103711. https://doi.org/10.1016/j.compgeo.2020.103711

Huang D, Gu DM, Song YX, Cen DF, Zeng B (2018) Towards a complete understanding of the triggering mechanism of a large reactivated landslide in the Three Gorges Reservoir. Eng Geol 238:36–51. https://doi.org/10.1016/j.enggeo.2018.03.008

Huang F, Huang J, Jiang S, Zhou C (2017a) Landslide displacement prediction based on multivariate chaotic model and extreme learning machine. Eng Geol 218:173–186. https://doi.org/10.1016/j.enggeo.2017.01.016

Huang J, Fenton G, Griffiths DV, Li D, Zhou C (2017b) On the efficient estimation of small failure probability in slopes. Landslides 14(2):491–498. https://doi.org/10.1007/s10346-016-0726-2

Ji J, Zhang CS, Gao YF, Kodikara J (2018) Effect of 2D spatial variability on slope reliability: a simplified FORM analysis. Geosci Front 9(6):1631–1638. https://doi.org/10.1016/j.gsf.2017.08.004

Ji S, Yu D, Shen C, Li W, Xu Q (2020) Landslide detection from an open satellite imagery and digital elevation model dataset using attention boosted convolutional neural networks. Landslides 17(6):1337–1352. https://doi.org/10.1007/s10346-020-01353-2

Jiang SH, Liu X, Huang J (2020) Non-intrusive reliability analysis of unsaturated embankment slopes accounting for spatial variabilities of soil hydraulic and shear strength parameters. Eng Comput (0123456789). https://doi.org/10.1007/s00366-020-01108-6

Ke G, Meng Q, Finley T, Wang T, Chen W, Ma W, Liu TY (2017) Lightgbm: a highly efficient gradient boosting decision tree. Advances in Neural Information Processing Systems. Morgan Kaufmann Publishers, San Mateo, CA, USA, pp 3146–3154

Kumar M, Samui P (2019) Reliability analysis of pile foundation using ELM and MARS. Geotech Geol Eng 37(4):3447–3457. https://doi.org/10.1007/s10706-018-00777-x

Li DQ, Jiang SH, Cao ZJ, Zhou W, Zhou CB, Zhang LM (2015) A multiple response-surface method for slope reliability analysis considering spatial variability of soil properties. Eng Geol 187:60–72. https://doi.org/10.1016/j.enggeo.2014.12.003

Li DQ, Xiao T, Cao ZJ, Zhou CB, Zhang LM (2016a) Enhancement of random finite element method in reliability analysis and risk assessment of soil slopes using Subset Simulation. Landslides 13(2):293–303. https://doi.org/10.1007/s10346-015-0569-2

Li DQ, Zheng D, Cao ZJ, Tang XS, Phoon KK (2016b) Response surface methods for slope reliability analysis: review and comparison. Eng Geol 203:3–14. https://doi.org/10.1016/j.enggeo.2015.09.003

Li S, Xu Q, Tang M, Iqbal J, Liu J, Zhu X, Liu F, Zhu D (2019) Characterizing the spatial distribution and fundamental controls of landslides in the Three Gorges Reservoir, China. Bull Eng Geol Environ 78(6):4275–4290. https://doi.org/10.1007/s10064-018-1404-5

Liao K, Wu Y, Miao F, Li L, Xue Y (2021) Time-varying reliability analysis of Majiagou landslide based on weakening of hydro-fluctuation belt under wetting-drying cycles. Landslides 18(1):267–280. https://doi.org/10.1007/s10346-020-01496-2

Liao W, Ji J (2021) Time-dependent reliability analysis of rainfall-induced shallow landslides considering spatial variability of soil permeability. Comput Geotech 129:103903. https://doi.org/10.1016/j.compgeo.2020.103903

Ling C, Lu Z, Zhu X (2019) Efficient methods by active learning Kriging coupled with variance reduction based sampling methods for time-dependent failure probability. Reliab Eng Syst Saf 188:23–35. https://doi.org/10.1016/j.ress.2019.03.004

Ling Q, Zhang Q, Wei Y, Kong L, Zhu L (2021) Slope reliability evaluation based on multi-objective grey wolf optimization-multi-kernel-based extreme learning machine agent model. Bull Eng Geol Environ 80:2011–2024. https://doi.org/10.1007/s10064-020-02090-5

Liu LL, Cheng YM (2016) Efficient system reliability analysis of soil slopes using multivariate adaptive regression splines-based Monte Carlo simulation. Comput Geotech 79:41–54. https://doi.org/10.1016/j.compgeo.2016.05.001

Liu LL, Zhang SH, Cheng YM, Liang L (2019) Advanced reliability analysis of slopes in spatially variable soils using multivariate adaptive regression splines. Geosci Front 10(2):671–682. https://doi.org/10.1016/j.gsf.2018.03.013

Liu Y, Zhang WG, Zhang L, Zhu ZR, Hu J, Wei H (2018) Probabilistic stability analyses of undrained slopes by 3D random fields and finite element methods. Geosci Front 9(6):1657–1664. https://doi.org/10.1016/j.gsf.2017.09.003

Massaoudi M, Refaat SS, Chihi I, Trabelsi M, Oueslati FS, Abu-Rub H (2021) A novel stacked generalization ensemble-based hybrid LGBM-XGB-MLP model for short-term load forecasting. Energy 214:118874. https://doi.org/10.1016/j.energy.2020.118874

Mualem Y (1976) A new model for predicting the hydraulic conductivity of unsaturated porous media. Water Resour Res 12(3):513–522. https://doi.org/10.1029/WR012i003p00513

Peng L, Xu S, Hou J, Peng J (2015) Quantitative risk analysis for landslides: the case of the Three Gorges area, China. Landslides 12(5):943–960. https://doi.org/10.1007/s10346-014-0518-5

Phoon KK, Kulhawy FH (1999) Characterization of geotechnical variability. Can Geotech J 36(4):612–624. https://doi.org/10.1139/t99-038

Song K, Wang F, Yi Q, Lu S (2018) Landslide deformation behavior influenced by water level fluctuations of the Three Gorges Reservoir (China). Eng Geol 247:58–68. https://doi.org/10.1016/j.enggeo.2018.10.020

Straub D, Schneider R, Bismut E, Kim HJ (2020) Reliability analysis of deteriorating structural systems. Struct Saf 82:101877. https://doi.org/10.1016/j.strusafe.2019.101877

Tang H, Wasowski J, Juang CH (2019) Geohazards in the Three Gorges Reservoir Area, China—lessons learned from decades of research. Eng Geol 261:105267. https://doi.org/10.1016/j.enggeo.2019.105267

Taylor SJ, Letham B (2018) Forecasting at Scale. Am Stat 72(1):37–45. https://doi.org/10.1080/00031305.2017.1380080

van Genuchten MT (1980) A closed-form equation for predicting the hydraulic conductivity of unsaturated soils. Soil Sci Soc Am J 44(5):892–898. https://doi.org/10.2136/sssaj1980.03615995004400050002x

Wang L, Yin Y, Zhang Z, Huang B, Wei Y, Zhao P, Hu M (2019) Stability analysis of the Xinlu Village landslide (Chongqing, China) and the influence of rainfall. Landslides 16(10):1993–2004. https://doi.org/10.1007/s10346-019-01240-5

Wang L, Wu C, Gu X, Liu H, Mei G, Zhang W (2020a) Probabilistic stability analysis of earth dam slope under transient seepage using multivariate adaptive regression splines. Bull Eng Geol Environ 79(6):2763–2775. https://doi.org/10.1007/s10064-020-01730-0

Wang L, Wu C, Tang L, Zhang W, Lacasse S, Liu H, Gao L (2020b) Efficient reliability analysis of earth dam slope stability using extreme gradient boosting method. Acta Geotech 15(11):3135–3150. https://doi.org/10.1007/s11440-020-00962-4

Wang MX, Huang DR, Wang G, Li DQ (2020c) SS-XGBoost: a machine learning framework for predicting newmark sliding displacements of slopes. J Geotech Geoenviron Eng 146(9):04020074. https://doi.org/10.1061/(asce)gt.1943-5606.0002297

Wang Y, Cao ZJ, Li DQ (2016) Bayesian perspective on geotechnical variability and site characterization. Eng Geol 203:117–125. https://doi.org/10.1016/j.enggeo.2015.08.017

Wang ZZ, Goh SH (2021) Novel approach to efficient slope reliability analysis in spatially variable soils. Eng Geol 281(1):105989. https://doi.org/10.1016/j.enggeo.2020.105989

Wang ZZ, Xiao C, Goh SH, Deng MX (2021) Metamodel-based reliability analysis in spatially variable soils using convolutional neural networks. J Geotech Geoenviron Eng 147(3):04021003. https://doi.org/10.1061/(asce)gt.1943-5606.0002486

Wu Y, Miao F, Li L, Xie Y, Chang B (2017) Time-varying reliability analysis of Huangtupo Riverside No.2 Landslide in the Three Gorges Reservoir based on water-soil coupling. Eng Geol 226(2):267–276. https://doi.org/10.1016/j.enggeo.2017.06.016

Yang B, Yin K, Lacasse S, Liu Z (2019) Time series analysis and long short-term memory neural network to predict landslide displacement. Landslides 16(4):677–694. https://doi.org/10.1007/s10346-018-01127-x

Yin Y, Wang H, Gao Y, Li X (2010) Real-time monitoring and early warning of landslides at relocated Wushan Town, the Three Gorges Reservoir, China. Landslides 7(3):339–349. https://doi.org/10.1007/s10346-010-0220-1

Yuan J, Papaioannou I, Straub D (2019) Probabilistic failure analysis of infinite slopes under random rainfall processes and spatially variable soil. Georisk 13(1):20–33. https://doi.org/10.1080/17499518.2018.1489059

Zhang W, Tang L, Li H, Wang L, Cheng L, Zhou T, Chen X (2020) Probabilistic stability analysis of Bazimen landslide with monitored rainfall data and water level fluctuations in Three Gorges Reservoir, China. Front Struct Civ Eng 14(5):1247–1261. https://doi.org/10.1007/s11709-020-0655-y

Zhang W, Wu C, Zhong H, Li Y, Wang L (2021) Prediction of undrained shear strength using extreme gradient boosting and random forest based on Bayesian optimization. Geosci Front 12(1):469–477. https://doi.org/10.1016/j.gsf.2020.03.007

Zhou J, Qiu Y, Armaghani DJ, Zhang W, Li C, Zhu S, Tarinejad R (2021) Predicting TBM penetration rate in hard rock condition: A comparative study among six XGB-based metaheuristic techniques. Geosci Front 12(3):101091. https://doi.org/10.1016/j.gsf.2020.09.020

Chapter 11
Future Work Recommendation

Firstly, how to rationally select an appropriate ML algorithm among numerous candidates and determine the associated hyperparameters remains an open question. Until now, many open-source packages of ML algorithms are available from the Internet and can be downloaded freely for scientific research. This open and friendly environment allows researchers and engineers to address geotechnical-related problems using different ML algorithms. After choosing the ML algorithm, the determination of the associated hyperparameters is a primary concern, which has a significant influence on the predictive performance. More and more optimization strategies have been developed to search the optimal hyperparameters with high efficiency (e.g., Bayesian optimization). Each ML algorithm has its own merits and shortages, and thus, no single or particular model can be always regarded as the most appropriate one in solving geotechnical problems. Thus, it is advisable to choose an appropriate ML algorithm according to the computational efficiency, memory consumption, and prediction performance in practical slope engineering applications.

Secondly, the influence factors considered when performing landslide susceptibility modeling using machine learning techniques are not comprehensive. Landslide susceptibility assessment provides an effective approach to quantify the likelihood of landslides occurring in a given area through portraying the landslide susceptibility maps. The previous studies focused more on the geometric variables (e.g., height and gradient), topographic features (e.g., elevation and aspect), and geological conditions (e.g., lithological property) in the landslide susceptibility assessment. Actually, some uninvolved factors also pose significant influences on the landslide susceptibility, such as the intense rainfall and groundwater table changes. Thus, the relative importance of these uninvolved factors in the landslide susceptibility assessment warrants to be investigated in future studies. Besides, although the likelihood of landslides occurring in a given area obtained from the landslide susceptibility assessment is a major concern in the geotechnical and geological engineering community, the landslide consequence also plays a significant role in landslide disaster prevention and mitigation, which needs to be further investigated.

Z. Wengang et al., *Application of Machine Learning in Slope Stability Assessment*,
https://doi.org/10.1007/978-981-99-2756-2_11

Thirdly, how to make the best use of the available numerous time-series monitoring data for site characterization and geotechnical reliability analysis. It is well recognized that parameter randomization is a significant prerequisite in the reliability analysis for considering the underlying uncertainties of geotechnical parameters. However, most geotechnical reliability analysis methods tended to generate the random samples of geotechnical parameters from an assumed probability distribution (e.g., normal distribution or lognormal distribution) associated with the statics (e.g., means and standard deviations), which neglects the role of available time-series monitoring data (e.g., displacement and pore water pressure data) in the interpretation of geotechnical parameters. Although geotechnical parameter characterization using the limited measurement data under the Bayesian framework has gained great achievements in the past few decades, how to incorporate the available time-series monitoring data for further updating the geotechnical parameters remains a challenge. Furthermore, most of the studies about geotechnical reliability analysis concentrated more on time-invariant reliability analysis (or static reliability analysis), and the time-variant reliability analysis (or dynamic reliability analysis) has been rarely reported. This may be attributed to the extensive computational efforts underlying the time-variant reliability analysis, and ML may open up a promising opportunity to facilitate the time-variant reliability analysis of geotechnical structures.

The last issue aims at the complex process of landslides and its occurrence mechanism, as well as the massive multi-source heterogeneous, multi-dimensional, multi-mode, multi-scale data, the traditional forecasting uncertainty based on numerical model and data assimilation is high, and the timeliness is poor, the big data-driven ML technique is applied to the relevant research, and the following research topics are encouraged: (1) development of multi-source heterogeneous sparse information mining and reconstruction methods for slope engineering stability assessment; (2) research on intelligent prediction and forecasting model, including the development of interpretable ML models to explore the intercorrelated characteristics and interconnections of catastrophic evolution contained in data, and lightweight engineering slope response prediction and forecasting; (3) research and development of data-driven prediction and forecasting physical model parameter improvement and model revision techniques, research on physical model and intelligent model integration and synergy methods, such as the physics-informed neural network, to establish a new data-knowledge-dual-driven prediction and forecasting system and a multi-driven cross-scale spatial–temporal response prediction method.

Appendix

Source Codes Used for Case Applications in Chapters

Chapter 5:

```
import pandas as pd
import numpy as np
from sklearn.tree import DecisionTreeClassifier
from sklearn.ensemble import RandomForestClassifier
from sklearn.model_selection import train_test_split
from sklearn.model_selection import StratifiedShuffleSplit
from sklearn.model_selection import cross_val_score
from sklearn.model_selection import GridSearchCV
import matplotlib.pyplot as plt
import time
from sklearn import metrics
from sklearn.utils import shuffle
from sklearn.metrics import confusion_matrix
import seaborn as sns
import xgboost as xgb
import sklearn.linear_model as lm
from sklearn import svm
from xgboost import plot_importance
from sklearn.metrics import mean_squared_error
import xgboost as xgb
import joblib
data0_df = pd.read_csv('滑坡点分区.csv').
data1_df = pd.read_csv('总滑坡点分区.csv').
def shuf(i,n):
i=shuffle(i[['water_distance','profile_curvature','plane_
curvature','aspect','slope','lithology',
'elevation','HAILS','syncline_distance','anticline_
distance','road_distance','NDVI', 'relief_amplitude','landuse',
'population','TWI','aridity','im','average_temp','average_
pre','label']],random_state=67).dropna().reset_
index().drop(columns=['index'])[:len(n)]
return i
data0_df1=shuf(data0_df,data1_df)
data1_df1=data1_df1[['water_distance','profile_
curvature','plane_curvature','aspect','slope','lithology',
'elevation','HAILS','syncline_distance','anticline_
distance','road_distance','NDVI','relief_amplitude','landuse',
```


Z. Wengang et al., *Application of Machine Learning in Slope Stability Assessment*,
https://doi.org/10.1007/978-981-99-2756-2

```
'population','TWI','aridity','im','average_temp','average_
pre','label']].dropna()
data_df=data0_df1.append(data1_df1)
data_df=shuffle(data_df).reset_index().drop(columns=['index'])
data=data_df.iloc[:,:-1]
label=data_df.iloc[:,-1]
feat_labels=data_df.columns[:-1]
Xtrain, Xtest, Ytrain, Ytest = train_test_split(data,label,test_
size=0.2)
params={'n_estimators':range(10,200),
'max_depth':range(1,30)}
rfc = RandomForestClassifier()
rfc_gc=GridSearchCV(rfc,params,cv=5)
rfc_gc=rfc_gc.fit(Xtrain,Ytrain)
print('Best: %f using %s' % (rfc_gc.best_score_, rfc_gc.best_
params_)).
rfc_best=rfc_gc.best_estimator_
score_train=rfc_best.score(Xtrain,Ytrain)
score_test=rfc_best.score(Xtest,Ytest)
print('best train score: %f, best test score: %f' %(score_
train,score_test))
Ypred=rfc_best.predict(Xtest)
Ypred_train=rfc_best.predict(Xtrain)
importances=rfc_best.feature_importances_
indices=np.argsort(importances)[::-1]
for f in range(Xtrain.shape[1]):
print("%2d) %-*s %f" % (f + 1, 30, feat_labels[indices[f]], impor-
tances[indices[f]]))
scorea=rfc_best.predict_proba(Xtest)[:,1]
import warnings;warnings.filterwarnings('ignore')
fpra, tpra, thresholds = metrics.roc_curve(Ytest, scorea)
roc_auca = metrics.auc(fpra, tpra)
plt.figure(figsize=(6,6))
plt.title('Validation ROC')
plt.plot(fpra, tpra, 'b', label = 'Val AUC = %0.3f' % roc_auca)
plt.legend(loc = 'lower right')
plt.plot([0, 1], [0, 1],'r--')
plt.xlim([0, 1])
plt.ylim([0, 1])
plt.ylabel('True Positive Rate')
plt.xlabel('False Positive Rate')
plt.show()
```

Chapter 8:

```
import pandas as pd
import numpy as np
import xgboost as xgb
import lightgbm as lgb
import catboost as cgb
from sklearn.metrics import r2_score as R2
from sklearn.metrics import mean_squared_error as MSE
import shap
#XGBoost
```

```
xgb_param = {
'objective':'reg:squarederror' # gamma tweedie mapeloss squared-
error
# ,'subsample':0.6
# ,'lambda':4
# ,'alpha':0.01207
## ,'max_depth':11
# ,"eta":0.01459
,'tree_method': 'gpu_hist'
,'colsample_bytree': 0.8616403624528273
,'gamma': 0.003490879281515258
,'learning_rate': 0.01604747710631525
,'max_depth': int(9.829019247369558)
,'min_child_weight': int(19.67062166576793)
,'subsample': 0.8294655849710145
}
num_round = 500 #520
#使用类Dmatrix读取数据
dtrain = xgb.DMatrix(Xtrain,Ytrain)
dtest = xgb.DMatrix(Xtest,Ytest)
bst = xgb.train(xgb_param, dtrain, num_round)
xgb_Ytrain_p = bst.predict(dtrain)
xgb_Ytest_p = bst.predict(dtest)
xgb_train_R2 = R2(Ytrain,xgb_Ytrain_p)
xgb_train_mse = MSE(Ytrain,xgb_Ytrain_p)
xgb_train_RMSE = np.sqrt(xgb_train_mse)
xgb_test_R2 = R2(Ytest,xgb_Ytest_p)
xgb_test_mse = MSE(Ytest,xgb_Ytest_p)
xgb_test_RMSE = np.sqrt(xgb_test_mse)
print('xgb_train_R2:')
print(xgb_train_R2,'\n')
print('xgb_test_R2:')
print(xgb_test_R2,'\n')
print('xgb_train_RMSE:')
print(xgb_train_RMSE,'\n')
print('xgb_test_RMSE:')
print(xgb_test_RMSE,'\n')
lgb_params = {'application':'rmse'
# ,'max_depth':11
# # ,'num_leaves':18
# # ,'min_data_in_leaf':4
# # ,'bagging_fraction':0.8932154969983767
# ,'learning_rate':0.013492964559329282
,'bagging_fraction':0.9439959921287611
,'feature_fraction':0.8993221903104023
,'learning_rate':0.02689374563971985
,'max_depth':int(19.663562226109295)
,'min_child_weight':int(22.607323142409612)
,'min_split_gain':0.00197977154575963
,'num_leaves':int(47.660797638718115)
,'bagging_fraction':0.9439959921287611
,'feature_fraction':0.8993221903104023
,'learning_rate':0.02689374563971985
,'max_depth':int(19.663562226109295)
```

```
,'min_child_weight':int(22.607323142409612)
,'min_split_gain':0.00197977154575963
,'num_leaves':int(47.660797638718115)
num_iterations = 5000 #1040
#读取数据
lgb_dtrain = lgb.Dataset(Xtrain,Ytrain)
lgbm = lgb.train(lgb_params, lgb_dtrain ,num_iterations)
lgb_Ytrain_p = lgbm.predict(Xtrain)
lgb_Ytest_p = lgbm.predict(Xtest)
lgb_train_R2 = R2(Ytrain,lgb_Ytrain_p)
lgb_train_mse = MSE(Ytrain,lgb_Ytrain_p)
lgb_train_RMSE = np.sqrt(lgb_train_mse)
lgb_test_R2 = R2(Ytest,lgb_Ytest_p)
lgb_test_mse = MSE(Ytest,lgb_Ytest_p)
lgb_test_RMSE = np.sqrt(lgb_test_mse)
print('lgb_train_R2:')
print(lgb_train_R2,'\n')
print('lgb_test_R2:')
print(lgb_test_R2,'\n')
print('lgb_train_RMSE:')
print(lgb_train_RMSE,'\n')
print('lgb_test_RMSE:')
print(lgb_test_RMSE,'\n')
#Catboots
from catboost import CatBoostRegressor
cgb_params = { "iterations": 2000
,"loss_function": "RMSE"
,"verbose": False
, "depth":3
, "bagging_temperature": 1.991
, "learning_rate": 0.07999
, "l2_leaf_reg" : 0.05393
}
cat_feat = []
cv_dataset = cgb.Pool(data=Xtrain,
label=Ytrain,
ncat_features=cat_feat)
#读取数据
cgb_dtrain = cgb.Dataset(Xtrain,Ytrain)
cgb   =   CatBoostRegressor(iterations   =   2000,   **cgb_params,
verbose=False).fit(Xtrain, Ytrain)
cgb_Ytrain_p = cgb.predict(Xtrain)
cgb_Ytest_p = cgb.predict(Xtest)
cgb_train_R2 = R2(Ytrain,cgb_Ytrain_p)
cgb_train_mse = MSE(Ytrain,cgb_Ytrain_p)
cgb_train_RMSE = np.sqrt(cgb_train_mse)
cgb_test_R2 = R2(Ytest,cgb_Ytest_p)
cgb_test_mse = MSE(Ytest,cgb_Ytest_p)
cgb_test_RMSE = np.sqrt(cgb_test_mse)
print('cgb_train_R2:')
print(cgb_train_R2,'\n')
print('cgb_test_R2:')
print(cgb_test_R2,'\n')
print('cgb_train_RMSE:')
```

```
print(cgb_train_RMSE,'\n')
print('cgb_test_RMSE:')
print(cgb_test_RMSE,'\n')
#SHAP
shap.initjs()
explainer = shap.TreeExplainer(lgbm)
shap_values = explainer.shap_values(Xtrain)
# visualize the first prediction's explanation
shap.force_plot(explainer.expected_value,     shap_values[0,:],
Xtrain.iloc[0,:])
shap.initjs()
shap.force_plot(explainer.expected_value, shap_values, Xtrain)
shap.initjs()
shap.dependence_plot("c", shap_values, Xtrain)
plt.figure(figsize=(12,10),dpi= 60)
shap.initjs()
shap.summary_plot(shap_values, Xtrain,show=False)
plt.figure(figsize=(12,10),dpi= 60)
# plot the global importance of each feature
shap.initjs()
shap.summary_plot(shap_values,            Xtrain,            plot_
type="bar",show=False)
```